BRIDGE BUILDERS

BRIDGE BUILDERS

MARTIN PEARCE AND RICHARD JOBSON

WILEY-ACADEMY

Illustration Credits

Every effort has been made to locate sources and credit material but in any cases where this has not been possible our apologies are extended. Cover, © Graeme Peacock/www.graeme-peacock.com; Frontispiece, © Richard Bryant/www.arcaid.co.uk; All images on pp.148-9 © Feichtinger Architectes. All drawings on pp.150-3 © Schulitz and Partners Architects, and photos © H Schulitz. All other drawings are courtesy of the first-named architects or engineers, as are all photographs except the following: p.28, © Rebecca May; pp.30 & 34-5, © Nicholas Kane/Arcaid; pp.40-5, © Richard Davies; pp.48 & 46, 48, 49 & 51, photos by Chris Gascoigne; p.50, photo by Cloud 9/Bob Lowrie; pp.52 & 54, photos by Martin Farquharson; p.56, photo by Geoff Warn; p.57, photo by Leon Bird; pp.62 & 64-5, photos by Nigel Young, © Foster and Partners; pp.66, 68 & 70, photos by Antonio Garbasso; p.72, photo by Paul Ott, Graz; pp.74-5 & 76, photo by Helmut Tezak, Graz; pp.78, 80-3, 178, 181 & 183, © Peter Cook/View; pp.84 & 86-7, © David H Davison/Davison & Associates, Dublin; 88 & 90-1, photos by Jocelyne van den Bossche; pp.116-9, renderings © Hayes Davidson/Nick Wood; p.120, rendering © Lifschutz Davidson/Mandy Bates; pp.122 & 124-7, photos by Marcus Robinson; pp.128 & 130-1, photos by J M Monthiers; pp.132 & 134-5, photos by Ian Smith; pp.136-7, © Napper Architects (Christopher Rainsford); p.138, photos by Bryan Wintermeyer, courtesy of Nicholas Lacey & Partners; p.139, © Mark Fiennes; pp.140 & 142-3, © Hans Werlemann/Hectic Pictures; p.144, © Shinkenchiku-sha/The Japan Architect co.; pp.146-7, © Timothy Hursley; pp.154 & 156-9, © Ana Maksimiuk for RFR; pp.161-5, photos by H G Esch, Hennef, courtesy of Schlaich Bergermann & Partner; pp.180 & 182, photos by Grant Smith; pp.184, 186-9, 190 & 192-3, 194 & 196-9, © Richard Bryant/www.arcaid.co.uk; pp.200 & 202-3, © Alan Williams; pp.204 & 206-9, © James Morris/Axiom; pp.210 & 213, renderings © Wilkinson Eyre Architects/Gifford & Partners; p.212, © Peter Mackinven; p.214, photo by Jayne Emsley, courtesy of Morning News and Gateshead Metropolitan Borough Council; p.215 (top), photo by Doug Hall & Lee Smith of Bonneys News Agency, courtesy of Gateshead Metropolitan Borough Council; pp.216, 218-9 & 221 (top), © Simon Warren.

Cover: Gateshead Millennium Bridge, Gateshead, England, Wilkinson Eyre Architects
Frontispiece: York Millennium Bridge, York, England, Whitby Bird and Partners

First published in Great Britain in 2002 by
WILEY-ACADEMY

A division of
JOHN WILEY & SONS
Baffins Lane
Chichester
West Sussex PO19 1UD

ISBN: 0-471-49786-X

Other Wiley Editorial Offices
New York • Weinheim • Brisbane • Singapore • Toronto

Design and Prepress: ARTMEDIA PRESS Ltd, London

Printed and bound in Italy

CONTENTS

The bridge swings across the stream 'with ease and power.' It does not just connect banks that are already there. The banks emerge as banks only as the bridge crosses the stream. The bridge designedly causes them to lie across from each other. One side is set off against the other by the bridge. Nor do the banks stretch along the stream as indifferent border strips of dry land. With the banks, the bridge brings to the stream the one and the other expanse of the landscape lying behind them. It brings stream and bank and land into each other's neighbourhood.

The bridge gathers the earth as landscape around the stream. Thus it guides and attends the stream through the meadows. Resting upright in the stream's bed, the bridge-piers bear the swing of the arches that leave the stream's waters to run their course. The waters may wander on quiet and gay, the sky's floods from storm or thaw may shoot past the piers in torrential waves – the bridge is ready for the sky's weather and its fickle nature.

Even where the bridge covers the stream, it holds its flow up to the sky by taking it for a moment under the vaulted gateway and then setting it free once more.

The bridge lets the stream run its course and at the same time grants the way to mortals so that they may come and go from shore to shore. Bridges lead in many ways. The city bridge leads from the precincts of the castle to the cathedral square; the river bridge near the country town brings wagons and horse teams to the surrounding villages. The old stone bridge's humble brook crossing gives to the harvest wagon its passage from the fields into the village and carries the lumber cart from the field path to the road.

The highway bridge is tied to the network of long-distance traffic, paced as calculated for maximum yield. Always and ever differently the bridge escorts the lingering and hastening ways of men to and fro, so that they may get to other banks and in the end, as mortals, to the other side.

Now in a high arch, now in a low, the bridge vaults over glen and stream – whether mortals keep in mind this vaulting of the bridge's course or forget that they, always themselves on their way to the last bridge, are actually striving to surmount all that is common and unsound in them in order to bring themselves before the haleness of the divinities. The bridge gathers, as a passage that crosses, before the divinities – whether we explicitly think of, and visibly give thanks for, their presence, as in the figure of the saint of the bridge, or whether that divine presence is obstructed or pushed wholly aside.

The bridge gathers to itself in its own way earth and sky, divinities and mortals.

Martin Heidegger: Poetry, Language, Thought, *translated and introduction by Albert Hofstadter (Harper & Row, New York; 1st ed 1975). Taken from a lecture given on August 5th 1951, 'Building Dwelling Thinking', in the course of the Darmstadt Colloquium II on 'Man and Space'.*

NEW BRIDGE AESTHETICS

Martin Pearce

It is with great discernment that Martin Heidegger, in his lecture on the relationship between man and space, chose to use the bridge as a metaphor through which to illustrate the central ideas of his philosophical thinking. He describes the bridge as a thing through which, for him, is revealed the nature of our consciousness and indeed the very essence of our being. Of the many objects of apprehension that surround and preoccupy us, why should it be that he chose the bridge as a conceptual vehicle through which to communicate his ideas? The answer to this lies in understanding the unique and complex nature of phenomena that we associate with the term *bridge*. This book represents an investigation of those phenomena, through the presentation of a range of recent structures that have had cause to address this most difficult question: what is a bridge?

The book evolved from the authors' interest in the wave of projects and competitions for bridge designs that arose towards the new millennium, and specifically pedestrian bridges. Collaborative teams of architects, engineers, and artists, along with landscape and environmental specialists, were now involved in the design of bridges, structures that historically had been the sole domain of the engineering profession. This joint approach to design brought into focus a realisation that in perhaps no other designed human artefact are the issues of technology – in the form of materials, structures, and construction – intertwined so closely with aesthetics.

Surveying the body of published material, we found that, with few exceptions, it generally dealt with bridges by categories of constructional typologies or structural methods. In addition, these texts largely focused on the structural properties of the bridge as stat-ically determinant systems, with quality evaluated in terms of structural efficiency and performance. In effect, the bridge was analysed as a calculable and mathematically quantifiable design product.

We realized that a complex range of issues, far beyond pure structural calculation, influenced the design of the new generation of bridges. From our different perspectives, one of us a teacher of design theory at the University of Portsmouth School of Architecture [Martin Pearce], the other a practising architect and design consultant at Whitby Bird & Partners engineers [Richard Jobson] we appreciated that the design of many recent structures was driven by far wider aesthetic concerns and a variety of complex approaches.

It also seemed apparent that the heroic large-span structures that had dominated the writing about bridges had been replaced by a more modest typology of pedestrian crossings. Foot and bicycle bridges had been designed and built towards the new millennium, partly as a result of the initiative to support more environmentally sustainable lifestyles through an enhanced infrastructure, and equally as a very public gesture by societies to mark the new millennium. No extensive survey had been undertaken of these structures. In our minds, this type of structure, perhaps more than any other, has provided an important vehicle through which to breakdown the differentiation between otherwise artificially distinct design professions. The nineteenth-century distinction between professional disciplines is perhaps nowhere more evident than in that most peculiar separation of design between architects and engineers. Only in recent years have these artificial boundaries come to be challenged, with interdisciplinary working producing some of the greatest built achievements at the end of the twentieth century. Certainly the high-tech modern architects have been quick to break with tradition, working in collaboration with some of the most innovative engineers to produce revolutionary buildings. However, the design of bridges has largely remained the domain of the engineering profession.

The new footbridges have evolved for two principal reasons. Firstly, the bridge is perceived as a powerful symbolic tool, marking and reflecting the growing sense of identity of regions, cities, and towns. Nowhere can this be seen more clearly than in the Wilkinson Eyre Gateshead Millennium Bridge. Secondly, as a unifying structure – connecting otherwise separate locations, and bridging across time – the bridge has become a powerful symbol of the new millennium. All over the world, structures have been built that carry the name 'millennium bridge', erected as monuments to the new era. Whilst many buildings, sculptures, and a plethora of other artistic works have been commissioned to mark the millennium, the building of new pedestrian bridges has been an unexpected, yet seemingly appropriate way to mark the event. Why should this be so?

This introduction to the bridges that follow seeks to address this question. Tracing the history of bridge design, it reveals how, through the ages, bridges provide powerful symbols of transition and change, and through this in some way capture an optimism for the future. It emerges that, in part, the appeal of the bridge relates to its very public nature. Primarily useful, the bridge provides a service to all who cross it, and is in all but a very few cases egalitarian in that it is open to all and presents no hierarchy of use or access.

Removed from its surroundings, the bridge is uniquely new. It jettisons the burden of history as new buildings or rebuilt parts of cities can not. As the architectural polemicist Lebbeus Woods has observed, in order to build on a site anew we often have to destroy that which once stood there, and thus most new buildings are erected on historic foundations. A new bridge, however, crosses virgin territory and presents little to which conservation lobbies can object, conveying a forward-looking message whilst being physically respectful through separation from its surroundings. Moreover, a new bridge brings with it the possibility of regenerating or bringing into focus its context in a manner that few other man-made constructions can achieve. Martin Heidegger's lecture cited the bridge as the means through which the banks of the river are brought to appearance 'as banks'. So too have many of these millennium bridges brought their locations to a new prominence, drawing popular attention to the unique qualities and possible needs of the local areas which they connect. In this way, bridges have been used as powerful catalysts for wider ambitions to regenerate areas. Whilst often far from obvious in their conception, once built these new structures seem to possess a strange inevitability. New physical connections and routes are quickly assimilated into the collective consciousness, becoming a familiar part of everyday life. It is perhaps for that reason that, in retrospect, it sometimes seems difficult to imagine a location without its bridge.

Forging physical connections between different areas is a powerful and practical way to bring people together. These connections come to symbolise unity and friendship, bridging geographical divides between different societies and cultures. Equally, as was the case with the Roman Empire, bridges can act as a means of control of one social group over another, and to some may come to represent tyranny. In contrast to their unifying nature, at times of conflict bridges can assume an even greater significance. The familiarity that makes us unable to imagine a town or city without its bridge is all too readily shaken in times of war, as the strategic importance of bridges to control and connect societies, their ideals, and religions is brought starkly into the foreground when bridges are destroyed. Burning bridges communicate a situation that once existed, and that may be lost beyond the possibilities of recovery. The act of destroying a bridge symbolises separation, isolation, defensiveness, and retreat from others.

These are situations through which the bridge shows its physical and cultural context in extreme and sometimes shocking ways. At these moments the bridge reveals itself as a peculiarly paradoxical object, being at once a means of connecting or dividing conceptual fields yet, at the same time, standing alone and singular as a complete artefact in itself. It is perhaps this inherent paradox that enables the particularly enigmatic qualities of the bridge to emerge, which have over the years provided a rich subject matter for many of the greatest painters. Raphael, Constable, Whistler, Cézanne, and Van Gogh repeatedly made representations of bridge structures. One sees in the constantly moving scattered reflections of the bridge in moving water a metaphor of their questions of representation, and the mirror of reality. Of these great painters, Monet's obsession with his garden bridge at Giverny in some way captures the objective of the Impressionists to understand the vivid and contrasting colours of the world around us, here reflected in the form of water lilies framed beneath the gentle arch of the simple structure. In this, Monet's search for the truth in painting is laid bare.

The closed geometry of a bridge's structure, combined with its unique location, make it more than many other structures identifiable as an iconic monument. Bridges throughout history have tended towards this hermetic singularity – Iron Bridge stands in stark contrast to its surroundings as do many of the structures that follow. As a result they exert a hold on our memory and imagination. Bridges, like great monuments, are icons and symbols of events and people, but unlike the statue of a great hero or a triumphal arch they are inherently useful, serving a public function, and they become a constant reminder in people's everyday lives in an immediate and physical manner.

It is not surprising then that great events, like the beginning of a new millennium, or great people, like George Washington, should have bridges named after them. However, the psychological importance of bridges has a considerable impact on the individual user – the act of crossing the bridge carries particular significance. It involves passing from one location to another, and, as with any journey, has a place of departure and the anticipation of eventual arrival. For the medieval monastic brotherhoods this journey was used as a quite literal analogy for our passage through life, describing the soul crossing the turbulent rivers of life from this mortal world into the kingdom of heaven. Chapels and chanteries located on bridge piers, such as at the Pont d'Avignon, served to celebrate and reinforce the message. Heidegger, too, viewed the bridge as a metaphor of life's journey, with each crossing being part of the inevitable road to the moment of death. Many cultures and societies the world over interpret bridge crossing in this way: the long section through the bridge reflecting the human life – growth from birth elevates us to the apex of our achievement in adulthood, followed by decline into old age and our arrival at our final resting place.

In quite literal terms, the bridge affords a new perspective on a particular, and perhaps familiar, location. It enables a location to be seen from a wholly new standpoint, and in so doing reveals something new of that place – from a position mid-span, for example, the banks present themselves in a particular and unique manner. Capitalizing on such a location, bridges such as that at York by Whitby Bird have incorporated seats and, combined with a curved plan form, provided discrete and identifiable places from which to observe the city.

The history of bridge design parallels the history of mankind's advances in technology. The new generation of bridges follows a similar pattern. Key to the development of many of the structures detailed in this book is the use of computer modelling and testing. The ability to build virtual models has enabled designers to consider bridges as fully integrated, holistic, structural systems. Testing the complete structure under various conditions would have previously presented a task of immense calculation time – using sophisticated computer systems, however, a variety of conditions and occurrences can be represented and evaluated. Through this process, the design of bridges has become increasingly pared down. For example, the inclusion of tuned mass dampers alleviates the need for even the slightest additional stiffness in the structure, as the dynamics of the bridge can be accurately predicted. Prior to computer modelling it was impossible to test such matters adequately, and bridges were inevitably over-designed to allow for a design process that could not predict such conditions. Today we see bridges constructed where the very limits of the structure have been tested, and elements refined to such a degree that even in its built reality we find ourselves questioning how, and indeed if, the structure will carry its load. The use of glass decks and the finest of suspension wires, such as that by Whitby Bird at the Science Museum, have created structures that are conceptually surreal; they appear as though they are unable to support themselves, and are sometimes a little frightening as they test our nerve and belief in technology. Bridge designers are now considering safety factors not only as a quantifiable phenomenon, but also as an important perceptual and psychological consideration for the bridge user.

In this way, the extremes of technology, both in the construction and in the design of bridges, have come full circle, with the most fundamental psychological human needs of comfort and safety outweighing the technical possibilities as an influence upon the final form. In contrast to the absolute computer-calculated limits of material strength and structural potential, attention to mankind's most primitive instincts and intuitive understanding of stability is exerting a significant influence on the form of modern bridge design today. To appreciate how this seeming paradox might come to exist, it is perhaps appropriate to return to the earliest beginnings and trace the history of bridge design.

THE BRIDGE IN HISTORY

The question of design does not perhaps, as such, relate to the first bridge used by humans. Our ancient ancestors' appropriation of ready-at-hand materials for use as tools and weapons no doubt extended to the convenience of passage afforded by the presence of a trunk of a fallen tree across rivers or other natural obstructions. Other examples of the use of naturally occurring materials might be imagined, such as the primitive suspension crossing formed from vines, or series of stepping stones or sequence of beam spans formed by piling rocks in a wide yet shallow stream. Little evidence survives of these earliest bridges, but their refinement in the form of developing timber structures and masonry arches is recorded in the accounts of the first civilizations in Mesopotamia by the Greek historian Herodotus, and from records of ancient China. It is not until the great classical civilizations of ancient Greece and Rome that we have direct evidence of the early bridges, and of these it is the Roman Empire that provides us with the first important structural achievements.

BRIDGING THE ROMANS

The success of the Roman Empire was founded on the construction of good communication routes. From the centre of power in Rome, radiating tendrils of the most comprehensive road network ever seen spread to the furthest outreaches of modern day Europe and Northern Africa. The construction of roads and their ability to link one town directly to the next meant overcoming natural obstructions. Fundamental to the success of the Empire was the need for these routes to be fast, efficient, and very reliable. Where roads encountered rivers, boat or ferry crossing was time-consuming as it involved the loading and unloading of goods and men. In addition, boat crossings were vulnerable to the vagaries of the stream. In contrast, bridges secured the permanence of communications that the Empire required.

Early Roman bridges were built of timber and there are some accounts of pontoon bridges. Timber performs well with respect to its strength to weight ratio, and is able to withstand a certain amount of bending. Simply supported timber beams could be combined to form reasonably effective structures for bridging. However, a stable and robust State needed similar structures, and the Romans achieved this through the use of that most durable of materials, stone. In contrast to timber, stone is heavy, limited to very short spans in the form of beams or lintels as self-weight becomes increasingly significant over the loads to be carried. In addition, stone is an inflexible material – instead of bending it cracks, and when it does so it fails immediately, often with disastrous consequences. But stone is permanent, and is not susceptible to decay or fire. Having selected this material for building empires, the question which faced the Roman engineers was how to employ the material in a useful manner.

This was resolved by the Romans' understanding of the properties of stone when used in compression. It was difficult to crush, and the form which best utilised this characteristic was the arch. The stone arch and its use in bridge building was one of the most important engineering advances ever, and became the form upon which, literally, the splendour of the Roman Empire was built.

The Roman stone arch has become such a familiar and ubi-quitous form that it is hard to appreciate it as an invention of considerable ingenuity. For the arch to work most efficiently and with maximum strength, each stone must endure maximum compression. For this reason, reducing the use of any compressible packing material, such as mortar between the stones, increases the strength of the bridge. The weight of the stone becomes crucial, since the load increases the strength of the structure by making each block hold more firmly against the next under friction. Indeed, many of the longest surviving stone arch bridges use no mortar at all and hold together under their own weight. With each wedge-shaped block (or *voussoir*) needing to press firmly against its neighbour, each stone had to be carefully cut and fashioned.

Whilst heavy materials pushing tightly together is an attribute for the completed structure, it presents a major difficulty when it comes to constructing the arch. The arch will only support itself once all the stones are in place and forced together. For this reason the Romans developed a method whereby the stones were supported by a timber structure known variously as *shuttering*, *centering*, *formwork* or *falsework*. Not only did the formwork need to be capable of holding the stones in place until the arch was complete, but it also had to allow for each of the stones to come under pressure, in a uniform manner, when it was removed. The formwork was often required to be held in place high above the ground, and for this springing stones were employed from which to build the formwork on piers. To appreciate a stone arch is to understand the process through which it was constructed.

The Romans were masters not so much of diversity in their inventions, but rather the perfection of a relatively few good ideas through extensive repetition and refinement. Having developed the skills to build arch stone structures, they used them everywhere that they could be worked to an advantage. Today, many structures remain with this use of the arch form, such the great Pont du Gard aqueduct (late first century BC or early first century AD) outside Nîmes in southern France.

The arch also provided a useful solution in the construction of large-scale building structures, where spanning characteristics could be employed to achieve a lightness and delicacy that monolithic structures would have been unable to attain. For example, the Pantheon in Rome (118–128 AD) is remembered for its coffered dome covering a vast circular space some 43.2 metres in diameter.

Supporting this dome is an equally impressive structural achievement. Taking the form of a series of linked arches, a cylindrical

Pont du Gard, Nîmes (*RIBA Library Photographic Collection*)

drum was built upon which the great dome rests. Here as elsewhere the structure – presenting a formidable engineering challenge in its own right – is ultimately covered by internal veneers of pilasters and niches. Imperceptible to the visitor, the supporting arch construction is only apparent high on the exterior of the building and is thus almost invisible from the surrounding narrow streets.

The stone arch provided a useful construction system where a single span could be made from one bank to the other. However, the limitations of the span meant that more than one arch was frequently needed, necessitating firm piers in the middle of the river. Often the surface material of the riverbed was unstable for building. The Roman solution again came from the development of a construction process, in the form of the *cofferdam*. This involved driving timber piles into the bed of the river to create an enclosure. For this the Romans developed pile-driving machines, which were floated out on barges into the middle of the stream. Often two concentric rings of piles were driven in, and then the space between in-filled with clay. This had the effect of forming a circular waterproof wall in the middle of the river, which, with the water bailed out, afforded dry working access to the riverbed. Work could then begin on the excavation of the riverbed and the construction of stone foundations from which the arches would spring. The principle of creating a structure that excludes water in order to undertake work on the riverbed has carried on into modern-day bridge pier construction. The development of prefabricated cofferdams (called *caissons*) has enabled the protective structure to sink down into the riverbed, rather like a pastry cutter, allowing much deeper excavation to be undertaken. Submersible pressurised caissons were developed in the nineteenth century to extend to very great depths where firm rock lay a long way below the water surface. However, in all of these developments the Roman principle of pier excavation remains key to the construction of bridges today.

Over time the shape of the foundation was developed to a point into the flow of the stream, so 'cutting' the water to either side of the pier. Later tapered on the downstream portion to become boat-shaped, the form of the pier was developed to further reduce downstream turbulence. Reducing the resistance of the piers to the force of the river led Roman bridge builders to another innovation. Rising water levels during floods created additional pressure on the piers and, with the semi-circular arch form, the higher the waters rose, the more masonry stood in the flow. To address this problem the Romans developed relief arches between the central springing of the main arches, which provided an additional escape route for flood waters. A series of these can be seen in one of the earliest Roman bridges, the Ponte Molle in Rome, which dates from the first century BC.

BRIDGING THE MEDIEVAL WORLD

Following the fall of the Roman Empire, Europe descended into the turmoil and uncertainty of the Dark Ages. During this period of about 700 years, little advance was made in respect to bridge building. Certainly bridges constructed under the Roman Empire continued to be used and maintained, but the social instability and lack of unified political structure lessened the will to forge any new permanent crossings. In England, the Norman Conquest marked a return to social order, and throughout Western Europe the establishment of new commercial trading links brought about a need for stable communication routes. Central to this new, more peaceful commercial order was the Christian Church. As civilisation emerged from the Dark Ages, the Christian Church became the principal source of social stability throughout Europe.

Ponte Molle, Rome (*RIBA Library Photographic Collection*)

Pont d'Avignon, France *(RIBA Library Photographic Collection)*

The bridge held a particular importance in the Christian faith. It had a powerful symbolic significance, acting as metaphor of the journey of the soul from this world to the next, crossing over from earth to heaven. It was also a practical symbol of Christian charity – as the church provided a place of sanctuary, so too did the bridge aid the traveller. However, beyond this benevolent and symbolic act of spreading faith, the building of bridges was also a pragmatic concern, connecting monastic communities and promoting and generating wealth for the Church.

The width of the medieval bridge carriageway was determined by the traffic to be carried. In most cases this was the width of one horse and cart, so many bridges of this time are no more than 2 metres at their narrowest point. Managing a single flow of horse-drawn traffic was possible over the length of the crossing, but with widths pared down to a minimum a passing cart and pedestrian traffic presented a hazard. The introduction of passing places, or *pedestrian refuges*, gave medieval bridges their familiar triangular piers. These refuges were often in effect the extension of another medieval bridge innovation, developed from Roman streamlining, the *cutwater* or arrow-shaped pier. This form offered less resistance to the flow of water, as it cut the water either side of the pier. The cutwater was also introduced on the downstream face of the piers, both to add additional buttressing to the structure and to make the pier into what was effectively a boat shape. This improved the laminar nature of the waters' passage, and lessened downstream turbulence. This practical solution to the passage of water also created places for carts and pedestrians to pass and in turn made bridges places to stop, talk, or meet.

The act of building places on bridges extended throughout the medieval period, and led to them becoming increasingly inhabited structures. Although buildings were erected on bridges throughout history, the construction of chapels and chantries had particular significance for the societies of the medieval period. Many bridges from the thirteenth century onwards had places of worship built on or above a pier on one side of the bridge. Of these the most famous is perhaps that at Pont d'Avignon (1177–85), crossing the river Rhône in south-east France. The bridge is not least remarkable in that it was built by a religious order by the name of the *Frères du Pont* (the sacred guild of bridge builders), created in the twelfth century with the principal aim of constructing permanent crossings.

The Pont d'Avignon is a superb example of a chapel built mid-way between the river banks, with all the powerful symbolic meaning that this carries. This idea was employed extensively throughout Europe, and many examples of bridge chapels survive – such as at Pernes-les-Fontaines, France; Calw, Germany; and Bradford-upon-Avon and Rotherham in England.

Developments in medieval cathedral building are today characterised by what is termed Gothic architecture, and many of the innovations in building technology found their way into bridge building of the time. The pointed or gothic arch, which had enabled the width of linear spans to be altered in cathedral building to form rectangular groin vaults, was readily incorporated into the design of bridges. The virtue of having a pointed arch over the Roman semi-circle was the ability to maintain a uniform springing and apex height of each arch, whilst varying the width of each span, thus allowing for the frequently irregular occurrence of suitable bearing positions across the riverbed. The disadvantage of the pointed arch was the requirement of a greater number of piers in the river, and the additional material and loading formed by the increased size of the arch abutments. These defects were in part addressed through the development of the *segmental arch* (a segment of a circle drawn from a centre below the springing line), which combined the Roman semi-circular form with the

Pont Neuf, Paris *(RIBA Library Photographic Collection)*

heavier piers, affording a flatter deck relative to the amount of material employed whilst still allowing for unequal linear spans.

Through the rediscovery of Roman bridge building techniques, lost during the Dark Ages, the *Frères du Pont* and other monastic communities disseminated the skills and knowledge of bridge construction throughout the Christian world. Combined with new abilities in stone crafts and masonry, which were being perfected through the craft guilds in the construction of the great medieval cathedrals throughout Europe, this marked a new era in bridge building.

BRIDGING THE CITY

From the beginning of the second millennium, the growth of stable city states brought a new standard of bridge building not seen since the Roman Empire. Paris and London developed rapidly throughout from the twelfth century onwards, with the nucleus of their expansion centred on the ancient river crossing. Along with the major streets, city squares, and important public buildings, the city bridges took on a new significance as an embedded part of the urban fabric.

The interrelationship of the bridge with a dense urban grain was nowhere more apparent than in Paris between the twelfth and eighteenth centuries. The original river crossing and Roman plan of Paris had developed as a result of a mid-stream island that reduced the crossing of the Seine to two relatively short distances. Focused on the Île de la Cité, Paris developed as a series of concentric rings with the island as its centre. The original Roman plan, contained in the outer ring of defensive walls, was first bisected by the east-west flow of the Seine and then by the north-south axis of the main connecting road. At their crossing, the Île de la Cité was connected by four principal bridges: Pont au Change, Pont Nôtre Dame, Pont Saint-Michel, and the Petit Pont. High urban density, coupled with the fact that the city's geographical,

administrative, and religious focus was on an island, gave a special and unique significance to the bridge linking the Île de la Cité. These bridges were an integral part of the urban structure, and took on a similar appearance to the surrounding streets as they themselves became inhabited with houses and shops. These structures took the form of buildings along either edge of a central street. Facing into the street, the buildings turned their back on the river, with the effect that the experience of crossing from one bank to the other became a continuity of built form, the river hidden behind the facades of surrounding buildings. With only limited numbers of crossing points, these streets built on bridges were of particular commercial importance. Retail space on the bridges was at a premium with higher levels of passing trade, and each crossing tended towards a particular type of commercial activity.

In Paris, as elsewhere, bridges became part of the living city structure. However, the warren that was the medieval network of streets and bridges did little for effective communication. Following the end of the Hundred Years War and the expulsion of the English in 1453, the Renaissance re-planning undertaken by Duc de Sully for Henry IV proposed the construction of a new bridge at the upstream end of the Île de la Cité to alleviate congestion. The Pont Neuf (literally 'new bridge'), completed in 1609, made a more direct connection between the two banks and is today one of the oldest bridges in Paris. Taking the form of de Sully's linear boulevards, the bridge is modulated by the medieval invention of semi-circular pedestrian refuges above each pier.

In London, the need to create a substantial and permanent crossing to connect the northern and southern sides of the river Thames led to the construction of the now famous Old London Bridge. Conceived by the priest Peter of Colechurch in the twelfth century, the bridge was completed in 1209, and is now legendary as one of the first inhabited bridges. Its deterioration and final demolition is remembered today in the form of a children's nursery rhyme: 'London Bridge is Falling Down'. The bridge stood for over 600 years and underwent many modifications. Engravings of the time show a structure surmounted with an irregular accretion of buildings, houses, shops and a chapel as the bridge became an extension of the city's fabric and life. A rich variety of activities took place across its span and the buildings dated from various times, having been erected, altered, and rebuilt in the various styles of the passing eras. Old London Bridge was a microcosm of the city that it served. Indeed it had one section which could be raised to allow masted vessels to navigate the Thames, in effect becoming a bridge upon a bridge.

Although common throughout the Middle Ages, very few inhabited bridges survive today. Of those that do, the Ponte Vecchio in Florence, built in 1345 by Taddeo Gaddi at the beginning of the Italian Renaissance, has many similarities to Old London Bridge. Ponte Vecchio (meaning 'old bridge') crosses the Arno river in the form of a city street, lined with houses and shops on either side. The bridge connects the principle urban spaces of the Piazza Signoria and the Palazzo Pitti, and constitutes one of the principal routes through this great Renaissance city.

Of all the cities that demonstrate the complex interrelationship of urban fabric and water crossing, none is more famous as a city of bridges than Venice. Situated on a river delta, water transport made Venice the commercial capital of the medieval period, connecting European and Eastern trade. The network of islands upon which the city is founded incorporates a total of about four hundred and fifty bridges. In Venice the canals are the streets of the city, populated by water traffic. The most important of these is the Grand Canal, which is accordingly spanned by the most important bridges. Where the canal is at its narrowest, in the commercial district of Rialto, it is crossed by the Ponte di Rialto. Replacing a series of previous timber structures, the current bridge is the result of an extraordinary design competition. The competition is remarkable in that Michelangelo, Sansovino, and the great Renaissance architect Palladio put forward proposals. But it was won by the appropriately named Antonio da Ponte, and the bridge was completed in 1590. It takes the form of a single stone arch, surmounted by two covered arcades of shops lining a central roadway, leading to central portico structures. Alongside, a parallel configuration and stepped walkways link the canal's banks. The porticos afford a stopping place from which to enjoy the views offered by the bridge's location at a bend in the canal. Whilst some regard the bridge's architecture as less than remarkable, its location forms a visual hinge in the cityscape and has made it the subject of many paintings. The exquisite re-presentations by Canaletto and Guardi, for example, demonstrate the focal picturesque quality and visual significance of the bridge as a landmark within the city.

Of the many Venetian bridges, it is perhaps one of the very smallest that carries the greatest poetic charge. Linking the Doge's Palace with the prisons on the other side of the canal, the Ponte dei Sospiri was designed by Antonio Contino in 1560. Its name, meaning 'Bridge of Sighs', refers to the wailing of prisoners being taken to stand before State inquisitors. An elevated arch supports an enclosed passageway, lit only by cage-like fretted windows. Like the gondolas that pass beneath, the bridge has become a romantic symbol of the city, a particular irony for a bridge of incarceration.

Although da Ponte's design for the Rialto Bridge was chosen, Palladio's submission for the competition was to have perhaps

Ponte Vecchio, Florence (*RIBA Library Photographic Collection*)

an even wider impact than the winning design. In his *Quattro Libri dell'Architetura* of 1570, Palladio published an outline of Roman bridge design as well as his own entry for the Rialto Bridge. Equipped with these patterns, Anglo-Palladian architects reproduced modified versions of his Rialto design. Palladio's ideas were, and indeed still are, widely exported. Capability Brown and William Kent's work at Stowe, Buckinghamshire, includes a bridge folly which strikingly resembles Palladio's Rialto design. Built as part of a great landscape garden designed in the picturesque manner of the eighteenth century, this jewel-like structure finds itself transported from its intended context of the dense urban Venetian background into a gently rolling Arcadian landscape.

Palladio's influence extended to one of the last inhabited bridges to be built as an integral part of an urban development plan. This occurred with the development of Bath, England, as a spa town throughout the eighteenth century. Connecting the expanding city with the Pulteney family estate lying across the Avon River to the south, a bridge was commissioned from Robert Adam. Completed in 1773, Adam's design comprises three arches, supporting two rows of eleven shops with attic stores above, set either side of a street. As with da Ponte's Rialto Bridge, attention is drawn to the middle of the span by a central portico structure.

Since Pulteney Bridge, many proposals for inhabited bridges have been put forward but none built. Perhaps this has resulted from the modernist tendency towards single-use programming and spatial understanding that saw the city as comprising individual freestanding objects rather than a network of interconnected voids. For whatever reason, Sir Edwin Lutyens' proposed art gallery spanning the river Liffey in Dublin, Konstantin Stepanovich Melnikov's bridge garage for Paris, W.F.C. Holden's proposal for the reconstruction of Tower Bridge, along with many others, remain unbuilt.

Over recent years an increasing interest in the need to recognise our cities as a sustainable form of human dwelling has led to a re-evaluation of inhabited bridge structures. An exhibition at the Royal Academy of Fine Arts in London (*Living Bridges*, 1996) chronicled the history of inhabited bridges. Alongside the exhibition, a competition was held for the design of a new 'living' bridge over the Thames, attracting interest from some of the world's finest architects and engineers. Whilst they are as yet unbuilt, it seems only a matter of time before increasing land prices and a need to better connect cities for pedestrians will lead to a renaissance of inhabited bridges.

BRIDGING THE INDUSTRIAL REVOLUTION

Throughout its history the bridge provides a record of the availability of natural materials, and man's ability to exploit and fashion these at his service. The masonry arch and timber truss remained the dominant material for bridge construction up until the late eighteenth century. The advent of the Industrial Revolution in England marked the beginnings of a fundamental shift in bridge design. Firstly, the development of the new, strong materials of iron and steel radically changed the limits of structural design; secondly, these enhanced structural possibilities came directly into the service of new types of traffic and a different kind of connection.

Nowhere is this revolution in bridge design, and indeed the social and cultural changes of the time, symbolised more powerfully than with Iron Bridge at Coalbrookdale. Iron Bridge has become an icon of the Industrial Revolution. It is therefore something of a paradox that the bridge is by no means the first such structure to employ iron as its principal material, nor is the span of any great distance; in fact it is of a quite modest scale.

Ponte di Rialto, Venice (*RIBA Library Photographic Collection*)

As the River Severn meanders through the Shropshire countryside of central England, it passes through the small village of Coalbrookdale. Here iron was first smelted in large enough quantities to make it viable as a primary building material, and at the centre of the Industrial Revolution it was from this focal point that Iron Bridge derives its reputation.

The bridge was designed by Abraham Darby III and Thomas Farnolls Pritchard (although uncertainty exists as to original authorship). It is formed from five semicircular ribs supporting a 7-metre-wide roadway, and is a rather literal translation of timber and stone building techniques of the time. The construction methods to accompany the new material had not yet been developed, so instead of bolts, nuts, and washers, the bridge uses traditional timber jointing such as mortices, dovetails, wedges, and screws to join components. In its construction method, however, the bridge marks an important transition towards the machine age. Each of the cast components, although unsophisticated in their fixing, are remarkable in that they were manufactured away from the site of their erection, so in this they represent some of the first prefabricated components ever used in building. Comprising 800 separate castings, the bridge's 400 tonnes of iron were floated downriver and assembled on site in the relatively short period of three months. Thus repetition of manufacture was joined by another, equally important phenomenon of the Industrial Revolution – speed of construction, enabled by prefabrication. The bridge reminds us of this new separation of manufacture and construction by the inscription cast into the outer rib:

This Bridge Was Cast At Coalbrook Dale And Erected
In The Year MDCCLXXIX

This is a metal structure that, beyond any other, stands in such contrast to the picturesque Arcadian beauty of the valley that it crosses. In a moment, it captures the new order and the relationship between man and nature that was to come.

The large-scale use of strong materials such as iron and steel transformed the design of building and bridge structures right through the nineteenth century. New technology demanded an approach to design that was 'engineered' through an understanding of material property, and precise mathematical calculation of the performance of these structures rather than trial and error. These new methods brought about the establishment of the distinct design disciplines of structural and civil engineering. Concentrating on technical and economic constraints, these new professions derived beauty not from reference to historical styles or established orders, but rather an aesthetic elegance resulting from pragmatic, functional necessity, fitness to purpose, and economy of material and construction.

No one more embodies these new disciplines than the earliest of professional civil engineers, Thomas Telford. Indeed he was one of the founder members of the Institute of Civil Engineers established in 1818. The son of a shepherd, he trained as a stone mason. He gained his construction experience during the building projects in Edinburgh of the late eighteenth century, being on the whole self-taught. The many civil engineering projects executed under his direction included the building of the Caledonian Canal, along with other important canals, roads, harbours, and bridges. However, to many his masterpiece of engineering is the Menai Straits Bridge, completed in 1826.

With the Act of Union in 1801, where Ireland was given parliamentary representation at Westminster, Dublin and London became linked politically. Telford took on the task of improving the overland connection from London to Holyhead, and the building of what is now the A5 trunk road. At the northern end of the route stood the barrier of the Menai Straits, separating the Isle of Anglesey from the mainland. The difficult ferry crossing

Iron Bridge at Coalbrookdale (*RIBA Library Photographic Collection*)

Menai Straits Bridge (*RIBA Library Photographic Collection*)

of the Straits had hitherto made the neighbouring town of Caernarfon one of the principal ports to Ireland, and local opposition to the building of a fixed link delayed Telford's project. The bridge was however begun in 1819 to Telford's unique design. The use of suspension structures for the crossing of deep gorges was well known, and rope footbridges had been used throughout history. Telford's ingenuity came in the translation of the principle of a timber deck suspended on tensioned hemp ropes, into an iron rib roadway suspended from eighty chains built up from wrought iron bars. Borrowing from the technology used in the rigging of ships, Telford worked to develop and test the stability of the tension structure. The cables, anchored back into the sides of the gorge, were suspended from two 46-metre-high limestone towers built out into the water with a span of 174 metres. The deck is connected to the catenary curve of the chains by suspension rods from which it hangs above the waters below. The bridge is of supreme elegance and lightness. As the first such structure, the principle of its design would inform all large span suspension bridges to follow. Of these Isambard Kingdom Brunel's Clifton Suspension Bridge, designed in 1830 but only completed in 1863, is a fine example of the bridge typology that was to come to dominate long span structures of the following century.

As the success of the Roman Empire was founded on the road network, the new industrial age became synonymous with the invention of the steam locomotive and development of an integrated railway network. A new breed of civil engineers set about devising engineering solutions to the problems of linking towns and cities with railways. Unlike roads, serving pedestrian and horse-drawn traffic, railways needed to run at a reasonably level gradient. Railway engineers capitalised on the innovations in tunnel construction methods that Telford and others had developed for Britain's network of canals. Aqueducts, however, were only used for the most extreme and otherwise unresolveable situations, due to the inherent difficulties of supporting large masses of water overhead. Economy of constructions, coupled with the slow speed of canal boat traffic, had made the use of locks a more economic way to accommodate irregular topographies. Railways, however, could not use such time-consuming means to cope with changes in the vertical landform, and new viaducts were required to run straight and level, regardless of the terrain or natural obstructions. In response to this problem the new engineers exploited the structural properties of the very materials from which the locomotives and rail track were being manufactured. Through their use of iron and steel, bridge engineers conquered even the most seemingly insurmountable of natural barriers.

Of these great civil engineers, Robert Stephenson and Isambard Brunel were the heroes of the steam age. George Stephenson had developed his famous *Rocket* steam locomotive for a competition of 1829. The success of his design heralded the beginning of the railway age. George Stephenson had also built the first railway route, covering 22 miles to connect Stockton and Darlington. But it was his son Robert who took on the challenges of creating a national rail network. Robert Stephenson used the new wrought iron in his building of the bridges needed by the rapidly expanding rail network of the early nineteenth century. In contrast to cast iron, which is overly hard and brittle owing to its high carbon content, the development of wrought iron provided a superior material for bridge construction.

Using this new material Robert Stephenson built the Britannia Railway Bridge that crosses the Menai Straits close to Telford's suspension bridge. Stephenson's structure was one of the first metal box frame bridges. Supported on stone piers, the bridge was formed from two huge wrought iron hollow tubes through which ran the rail line. Robert Stephenson also made considerable advances in the process and speed of bridge construction through his development of the steam hammer, used for driving piles into the riverbed and avoiding dangerous and time-consuming construction of pier foundations with submersible caissons.

Among the great monuments to this age of steam are the vast railway sheds at King's Cross and Paddington, and the miles of tunnels such as the 3.5 mile tunnel (the Woodhead) of the Manchester–Sheffield Railroad (1839-45).

Gustave Eiffel was one of the first truly international design engineers, undertaking projects throughout many countries. His involvement with the construction of the Statue of Liberty, along with his famous tower in Paris, leave him responsible for two of the most powerful symbols of nationhood ever built. In addition to these projects, Eiffel built hundreds of bridges of different kinds, for which his design ability was equally matched by his business acumen, his own construction company executing the work and providing all the steel. Of his bridges the viaduct at Garabit, over the Truyère River in the Massif Central Region of France, demonstrates Eiffel's genius to build structures of immense scale yet extraordinary delicacy. Crossing a deep gorge, the rail tracks are supported 122 metres above the water. For many years this was the tallest bridge in the world. The bridge takes the form of a parabolic arch from which struts are raised to provide intermediate supports to a box-section beam carrying the tracks above. The arch spans 165 metres, but is of such a delicate skeletal form that its lightness of touch on the sides of the valley leaves the visual continuity of the landscape unaffected by its presence.

Eiffel is perhaps best known for his famous tower in Paris, built for the *Exposition Universelle* of 1889. The tower follows the

Forth Rail Bridge, Scotland (*RIBA Library Photographic Collection*)

principles of prefabrication and off-site factory manufacture of Iron Bridge. However, with Eiffel's tower, 1,200 wrought iron components fully exploit the potential of this approach to construction. One year after this great symbol of modernity and industrial production was completed, one of the other great engineering achievements of the nineteenth century was finished – the Forth Rail Bridge. Spanning the Firth of Forth to link the western side of Scotland, this viaduct showed structural daring and ingenuity on a hitherto unseen scale. Like the Eiffel Tower it is one of the first examples of a built form as pure structure. Sublime in scale and evoking emotions of awe it is uncompromising in its singularity of purpose and demonstrates mankind's ability, through the mastery of material and structure, to overcome the forces of nature.

Although the Forth Rail Bridge is an icon of engineering achievement, it was built in the shadow of a great engineering disaster. The railway from Edinburgh to Dundee crosses both the Firth of Forth and the Firth of Tay. The first bridge to be built on the route was that over the Tay. Designed by Thomas Bouch, the bridge made the 13.2-kilometre Tay crossing by means of a series of 89 wrought iron trusses supported on slender cast iron columns. The bridge had been opened for less than two years when, on a stormy December night in 1879, a catastrophic failure occurred just as a fully laden passenger train was crossing.

Faults in the iron casting, coupled with Bouch's failure to predict the effects of wind pressure on the bridge, led to 75 passengers losing their lives as the train plummeted into the freezing waters below. Bouch's career was ended, his design for the Forth crossing dropped, and the board of trade suspended the use of steel in bridge construction. It was only with the lifting of this ban, and extensive testing for wind loads, that the design of Benjamin Baker and John Fowler for the Forth Rail Bridge was approved.

Their design employs the use of huge cantilevers, which extend like outstretched arms from either side of the three circular caisson pier foundations. The cantilevers, the bases of which are 513 metres apart, are connected at their tips by suspended trusses bridging the remaining 107 metres from the end of one cantilever to the next. The lattice framework of the supports is built from huge steel tubes, some up to 3.6 metres in diameter, and the whole structure is riveted together. The bridge is a spectacular achievement of pure engineering, and broke several records for man-made structures at the time of its completion. Its sheer scale persists in the folklore surrounding the time it takes to be painted.

The Forth Rail Bridge marked the pinnacle of railway bridge building. As the twentieth century dawned, a new form of human communication was to have a profound effect on the nature of bridge building – the advent of the internal combustion engine and in particular the motor car.

BRIDGING AMERICA

The history of bridge building at the beginning of the twentieth century is not only dominated by that of the motor car, but also by the continent that most rapidly embraced the potential of powered road transport, North America.

The importance of bridges in the colonisation of the New World was well understood by the early settlers. The availability of natural materials dictated the construction of these early bridges. The dense forests of New England provided the raw materials, and the timber truss became the predominant structural form used to forge new routes out into the wilderness. Simple structures were quick to assemble and required only the framing skills that were used in timber housing of the time. Expedient in construction, such bridges however suffered from their vulnerability to fire, floods, and rot resulting from exposure of the timber to the elements.

Philadelphia covered bridge (*RIBA Library Photographic Collection*)

It had been long known that the life span of timber frame houses could be extended by over-cladding the primary structural elements. A pitched roof of timber shingles and an external wall with boarding are familiar as the 'clapboard' houses of New Hampshire. Translating these protective devices into the design of bridges resulted in what are now known as *covered bridges*. Reflecting the entrepreneurial society of the early United States of America, the construction of bridges was financed privately from newly formed companies and collectives, with investors deriving return from tolls levied on passing traffic. By covering the bridges they became useful as places of shelter and meeting, and were often used to display advertising hoardings. Indeed they are sometimes referred to as 'kissing bridges', as a place for secret assignations. However, the reality of the covered form is one of simple practicality – covered wooden supports last longer when protected from the elements. In time, and as new forms of transport and improved communication routes developed, many of the bridges fell into disrepair and today relatively few of these enigmatic structures survive.

With the coming of the railways, trussed timber bridges continued to provide one of the principal means for negotiating natural obstructions. Increased loads and vibration of steam locomotives and passenger wagons brought about the development of structures which incorporated high-tensile materials, such as steel, to increase the performance of the timber structure. Bolted timber connection and plate connection took the areas of high stress, and weight reduction through the replacement of tensile timber members with steel rods led to new structural forms such as the *bow string truss*.

However, motorised traffic brought about the need for a new kind of bridge and different engineering solutions. The use of thin steel rods with their high tensile strength had been employed in the construction of suspension bridges towards the end of the nineteenth century. By spinning the wires together, exactly as rope manufactures wove thin threads together to form strong ropes, extremely high-breaking strains could be achieved in the production of cables. The Brooklyn Bridge of 1883 was the first large-scale structure to employ spun steel cables. Designed by John A. Roebling and executed by his son, the bridge uses four steel cables, suspended between two granite towers and anchored into the ground at both the Brooklyn and the Manhattan banks. Each cable is made up of spun together thin steel strands to support the roadway below through a web-like network of secondary cables. At the time of its completion the bridge was the longest suspension structure in the world, and today stands as an enduring monument to the revolutionary impact of steel on bridge design.

Brooklyn Bridge, Williamsburg Bridge (1903), and Manhattan Bridge (1906) remained the great bridge structures of New York until completion of a steel bridge of a far greater scale. The George Washington Bridge across the Hudson River, completed in 1931, was designed by a brilliant engineer Othmar H. Ammann, who designed many of North America's great bridges. Commissioned to alleviate the extensive traffic congestion that the rapid expansion of the motor car had brought to Manhattan, the George Washington Bridge was originally designed to carry six lanes of motor traffic with four tramway lines planned on a second deck below. The lower deck was only added subsequently, and now both decks carry lanes of motor traffic, each in the opposite direction. Not only is the George Washington Bridge a masterpiece of steel engineering – Brooklyn Bridge spans 480 metres, while the George Washington Bridge covers over twice that distance at 1,060 metres – it is also a testament to the designer's foresight, that the bridge has been capable of expansion to accommodate the progressive increase of traffic.

CONCRETE BRIDGES

The Romans had first developed the use of water, sand, stone, and a binding agent to form concrete. Higher strengths, through the use of Portland cement and metal reinforcing, further advanced the usefulness of this material in bridge construction through the nineteenth and early twentieth centuries.

Combining the tensile strength of thin wire or cable with the compressive strength of concrete brought about another revolution in bridge design. First developed by the French Engineer Eugène Freyssinet, pre-stressed concrete differs from reinforced concrete as high-tensile steel wires are tensioned while construction takes place. This optimises the material structural performance by avoiding further stretching of the reinforcing when load is applied, and in so doing alleviates differential stresses between the concrete reinforcing. Today pre-stressed concrete is ubiquitous in road building, in the form of the underpasses and fly-overs that make up our road network. For a number of these structures road engineers have turned to Freyssinet's invention, and pre-stressed concrete road bridges have become a part of everyday experience. Highway engineers were quick to adopt the material for its economy of construction and flexibility in developing irregular forms. However, at the beginning of the twentieth century the use of concrete in the building of large-scale structures was new and revolutionary. Eugène Freyssinet did not come immediately to the pre-stressed concrete system. His experiments with concrete structures were built on research carried out on the use of steel as opposed to iron in concrete reinforcing, undertaken by François Hennebique in the late nineteenth century.

Together with Freyssinet's work, Swiss engineer Robert Maillart was equally influential in the development of concrete in bridge construction. Maillart's early work was relatively traditional in its form, and like Iron Bridge at Coalbrookdale before, his designs substituted the new material in historic forms. Maillart's unique contribution came when he broke new ground with his design for the Tavanasa Bridge over the Rhine in 1905. Fully exploiting the plastic qualities of concrete, Maillart cut away unnecessary material by trimming triangular shapes out of the web structure near to the abutments. Maillart went on to develop these plastic qualities to great effect in the form of a stiffened slab arch in his design for the Schwandbach Bridge in Switzerland, the bridge curving in plan as it springs across the valley gorge. Maillart was also responsible for perhaps one of the most powerful icons of modernity in bridge designs, that of the Salgina George Bridge of 1930, again in Switzerland. Spanning a modest 90 metres, yet set some 75 metres above the valley floor, this delicate structure glistens in its white perfection against the irregular, rugged landscape beyond.

The relatively modest structure being developed through Maillart and Freyssinet's pioneering work at the beginning of the twentieth century, as adopted for highway construction, would go on to have one of the most prominent impacts on landscapes the world over. However, Maillart and Freyssinet are not widely known. In contrast, at the time of their working, several of the greatest and most monumental of bridge structures were being erected. Sydney Harbour Bridge, completed two years after Salgina George Bridge, is today a symbol of that city and of the Australian continent. More than any other it is hard to imagine this particular city without its bridge and Opera House. Indeed the districts of the city are now organised in accordance with the bridge's location. The sense of the structure's importance is reinforced by its dramatic location at the entrance to Sydney harbour. Spanning 500 metres and designed by Sir Ralph Freeman, the bridge takes the form of a deck suspended from a great steel lattice arch, the springing points of which are marked by pairs of pylons. Tapering skywards like those of ancient Egypt, the pylons on either side perform no structural function. In fact the upper string of the arch stops short of the piers, but the pylons add a sense of stability in contrast to the lightness of the arch.

As Sydney Harbour Bridge had assured a major city a place in worldwide consciousness, so too did a bridge of very different design built only five years later. The Golden Gate Bridge in San Francisco has not only become synonymous with that city but also with suspension bridges the world over. As with Sydney, the bridge crosses the mouth of one of the world's great natural harbours. The deep waters of Sydney harbour had meant the arch was constructed as cantilevers from either bank eventually meeting mid-span. In the same way, the deep water channel between Marin County and San Francisco eliminated the possibility of constructing temporary structures in the channel during erection. Designed by Joseph B. Strauss, the bridge spans 1,280 metres. The deck is suspended from wires looped over the centenary suspension cables to form pared hangers, giving the impression of passing along a great transparent colonnaded street. The bridge is famously painted orange in colour to complement the burnt ochres of the surrounding landscapes. However, this choice of colour, intended to blend with its context, also results in the bridge's memorable visual prominence when viewed against the blue sea of sky. New York, San Francisco, and Sydney, as riverside cities, have come to be inextricably linked to, and in some way defined by, their bridges. Some of our greatest technological achievements are represented by these structures.

Mastering nature by crossing ever-greater obstacles has focused bridge design as a technological discipline, pushing the boundaries of materials and structural systems to achieve ever greater

Sydney Harbour Bridge, Australia *(RIBA Library Photographic Collection)*

spâns with yet more slender structures. In recent years a new generation of bridges has concentrated not simply on pushing technological possibilities. The new millennium has refocused our attention on bridges as symbolic structures in their own right, and this emphasis on inherent meaning has moved away from the monumental gesture of structures like the Golden Gate Bridge and their global statement. Instead bridges have come to be recognised for their local significance. As part of this development, whilst the infrastructure bridges that link together road and rail networks continue to be built, a newly found belief of the importance of sustainable ways of living has questioned the dominance of the motor car – giving greater priority to more environmentally friendly forms of transport and, in turn, to a new range of pedestrian and cycle bridges.

In tracing the history of bridges, this introduction explores the relationship between human needs and available materials, and the operational importance of bridges for societies and empires. In addition, it examines the psychological importance of bridges, the way in which they come to represent great events or people and become part of our collective memory.

The footbridges illustrated in this book not only make places and connections that enhance people's daily lives through their functional usefulness, but have also been designed to form places and events in their own right. In so doing they have been forced to address issues associated with unique locations and historical contexts, and from this many derive their quality or form. In line with this challenge, the boundaries between traditional design disciplines have been increasingly broken as engineers, architects, artists, lighting designers, and a host of other professionals have worked together in these most public of structures. This has resulted in a radical re-evaluation of the essential qualities that constitute good bridge design. The beginnings of this enquiry are manifest in the many beautiful structures that follow in this volume. These extraordinary pedestrian bridges mark the beginning of the now emerging concept of new bridge aesthetics.

WEST INDIA QUAY FLOATING BRIDGE

London, England

The objective of this bridge design was to span between two points with a structure as elegant and minimal as possible while conveying a sense of freedom and space to pedestrians as they crossed the water with unimpeded views of the Docklands. The bridge links two important areas of different scale – the large commercial development of Canary Wharf and the smaller West India Quay with its nineteenth-century warehouses.

By creating a floating structure, emphasis is placed on the stretch of water, the visual prominence being horizontal rather than competing with the vertical dominance of Canary Wharf. The concept of the pontoon bridge has been known since 2000 BC. This design reinterprets that ancient tradition by creating a gently arched structure floating on four pairs of pontoons.

All fabrication took place off-site in a factory, which ensured a high quality product. The bridge was then transported in two halves and floated into position from an assembly area at Royal Albert Dock. The platform tapers towards the centre, increasing the sense of perspective and the lightness of the structure. The deck is gently ramped on either side, joined by a central opening section, adding a graceful profile to the 94m length. The mid-span element is lifted hydraulically, with a simple balanced cantilever action allowing boats to pass through. The decking is aluminium, while the main body of the bridge is mild steel. At night stainless-steel handrails with integral lighting illuminate the bridge surface, emphasising its floating quality, and runway lights delineate the softly curved plan, changing colour as you pass by. This animates the entire stretch of water, providing dramatic views from the surrounding buildings and giving the dock area a strong visual identity.

KELVIN LINK BRIDGE

Glasgow, Scotland

This footbridge links Glasgow University with Kelvingrove Art Gallery across the River Kelvin. The bridge is 'S' shaped, and winds its way sinuously through Kelvingrove Park, the site of several exhibitions in Scotland at the turn of the twentieth century. Forming two kissing bridges, a secondary structure connects the principal link to the network of routes through the park.

The concept of forming a simple and elegant deck of brick paving is supported by a steel and concrete structure, in turn held up by a series of splayed, slender legs at regular intervals. Where the river is bridged, these splayed legs are picked up by two arches that transfer their forces into abutments at the riverbanks. Interconnected at the positions of the legs these arches form a stable system, the inner carrying the greater load, the outer providing the structures for a second bridge path.

The deck structure is asymmetric in cross-section, based on a tubular spine beam and tapering ribs formed out of steel sections, to which pre-cast concrete deck elements are fixed, so acting compositely with the spine beam.

The asymmetry of the structure emphasises the light and dynamic appearance of the bridge, and at the same time caters for the torsional effects resulting from the meandering form of the route.

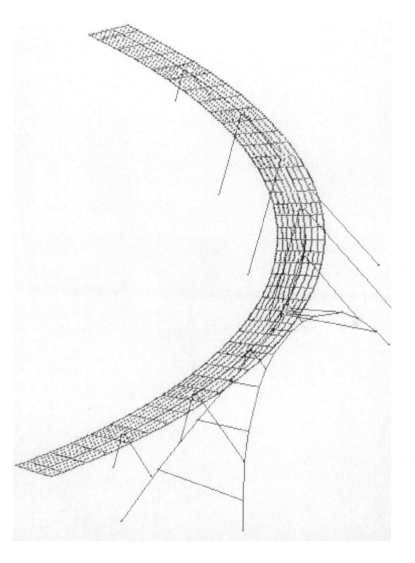

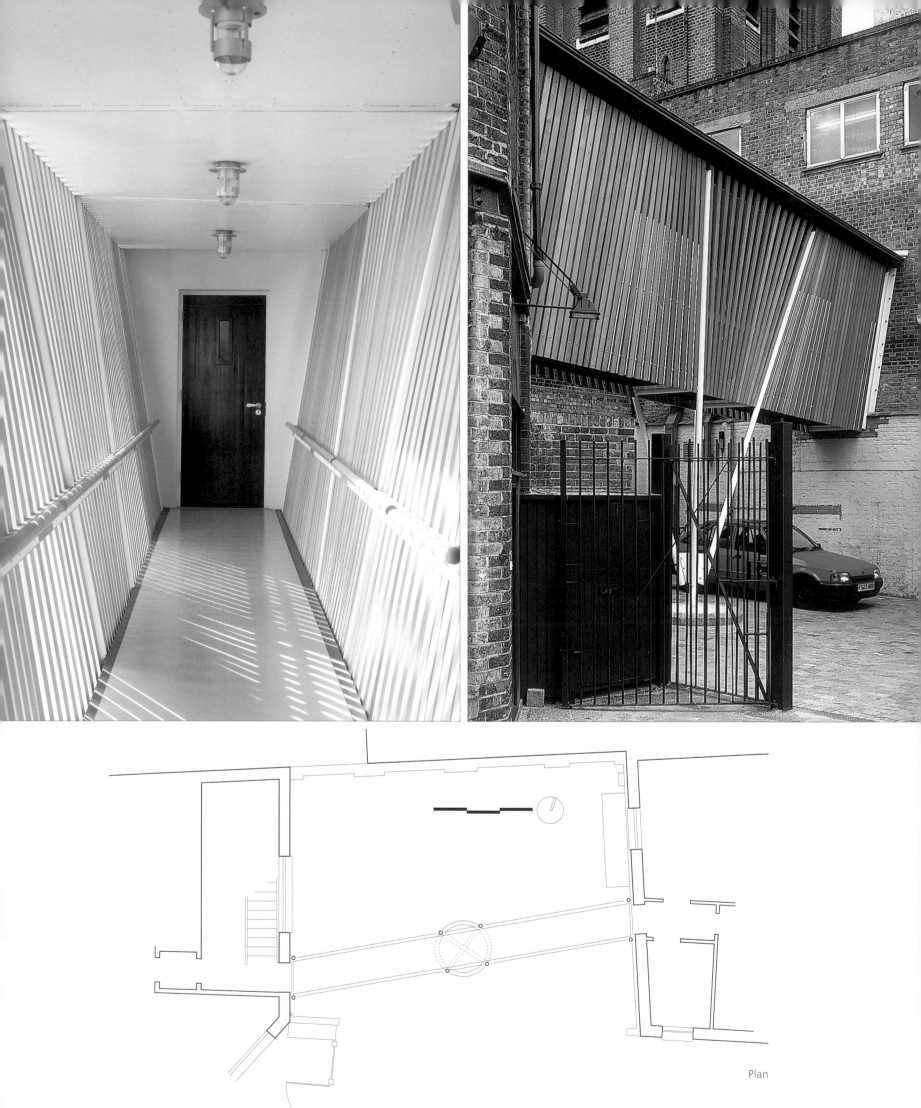

Plan

BRIDGE AT THE WORLD ASSOCIATION FOR CHRISTIAN COMMUNICATION

London, England

The World Association for Christian Communication owned two buildings, one fronting onto the street and the other at the rear, separated by a small car park. Forming a bridge at first floor level provided an innovative way to improve communication amongst the staff. In order to achieve this a number of features and requirements needed to be taken into account – the different levels between the two buildings; a skew in plan between locations where the bridge could be accessed; the retention of all car parking spaces; and the imposition of the least load possible on the existing walls in order to avoid underpinning or other reinforcement.

The solution was a semi-open bridge made of timber joists supporting floor and roof. In turn, these are supported by trapezoid steel frames, bolted to the existing brickwork at each end of the bridge, and framing the new doorways. The cladding to the sides of the bridge comprises western Red Cedar square-edge boards, set in place with caps of varying sizes, in order to create a splayed effect which, at each end of the bridge, allows the cladding to sit neatly against the walls. The interior of the bridge splays outwards giving the effect of a wider corridor, and leaving space for oak handrails down each side. The floor coverings are textured rubber, while the sub-floor and soffit are sheathed in plywood which provides lateral stability to the structure.

The simple act of moving between two buildings is enriched by elements such as the aroma of the cedar when it rains, the sound of rain on the roof, wind coming through the partial protection provided by the slats, and the rhythm of the light – all contributing to the sense of being partly inside and partly out.

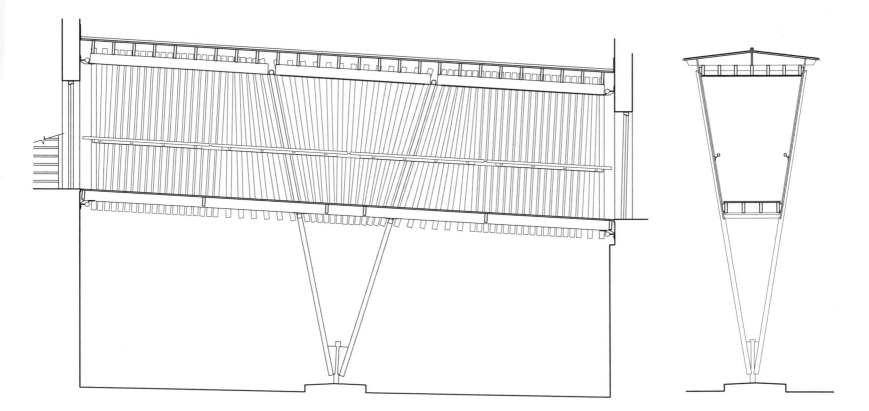

L-R Elevation and section

PLASHET SCHOOL FOOTBRIDGE

Newham, East London, England

Plashet School Footbridge was completed in September 2000, and provides a unifying link for pupils and staff over a busy road between the original school and the 1960s school building to the north. The footbridge utilises pioneering technology to create a dynamic structure, facilitating safe weather-protected connection and enriching the experience of crossing between school buildings. The structure was conceived to be extremely cost-effective and to minimise disruption to the school during construction.

The 67m long covered footbridge is a fluid form, curving past and preserving a mature tree in the school grounds. A seated viewing gallery at the mid-point above the road provides a rest spot for the children, with views to the outside world. The inexpensive steel structure minimised steel fabrication, and has a low cost, low maintenance, translucent Teflon fabric covering for weather protection and natural illumination.

The bridge is constructed of a steel carriage beam supported on steel columns, covered by a lightweight PTFE fabric stretched over a series of galvanised steel hoops. The carriage beam is formed from standard 915m

high universal beams, with a steel floor deck welded between. Circular steel tubes on top of the beams complete the balustrade.

Detailing includes prominent steel plate welding to echo Newham's shipbuilding history. The columns are of steel plate cut to a silhouette of outstretched palms. Each end of the footbridge is supported on pairs of smaller silhouetted columns, which splay out to frame the north school entrance. The bridge is clamped to the two central columns by wedges driven through large hollow pins. In contrast the galvanised lightweight steel elements, the hoops, drainage hoppers, and gargoyles bolt on to the carriage structure.

Rainwater collection is expressively detailed to provide an animated feature on rainy days. Collected by fabric gutters, the rainwater is discharged via galvanised steel hoppers into the tubular steel handrails. These descend to the gargoyles, which spout the rainwater via chains down the column face to gullies at ground level. Internal lighting is provided by torch lights, mounted in the round apertures formed at the apex of the steel hoops.

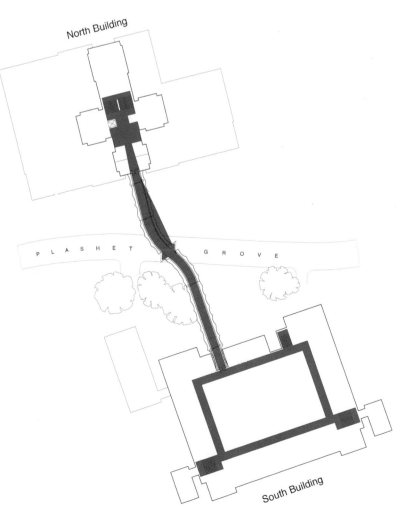

North Building

PLASHET GROVE

South Building

Site plan

Right Viewing gallery section
Centre Columns elevations
Below Isometric kit parts,
Isometric drainage detail

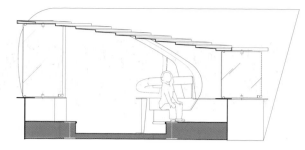

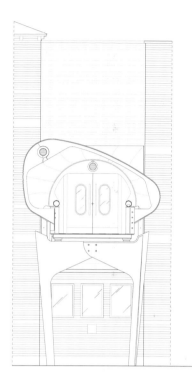

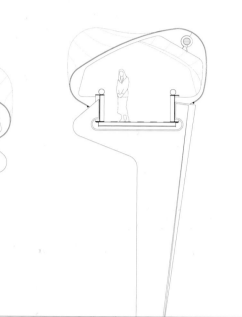

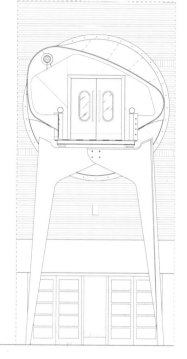

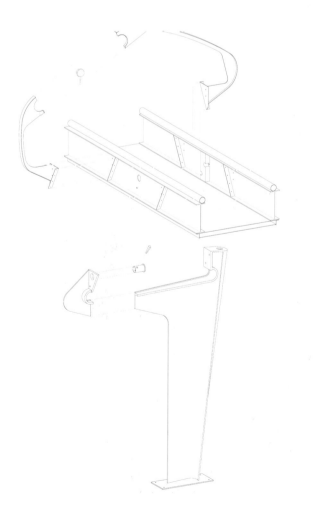

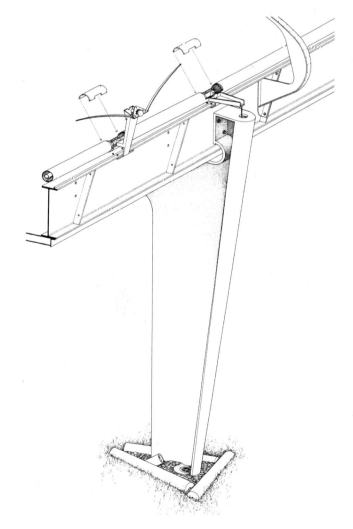

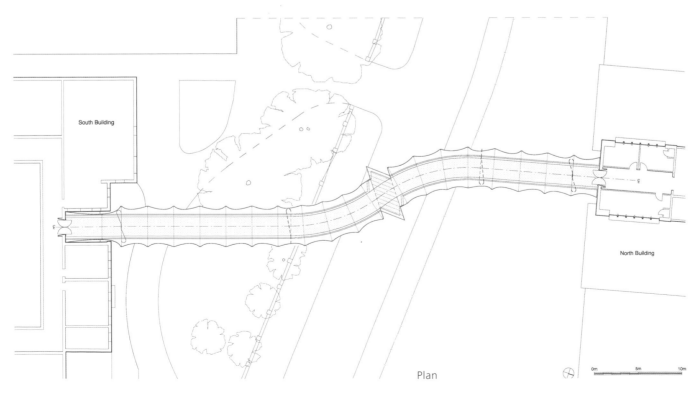

South Building

North Building

Plan

0m   5m   10m

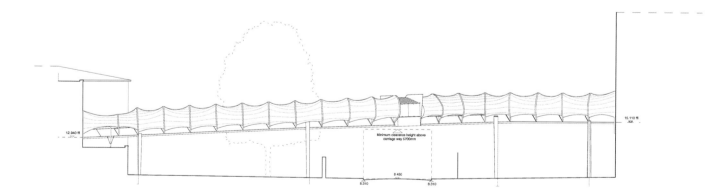

Elevation

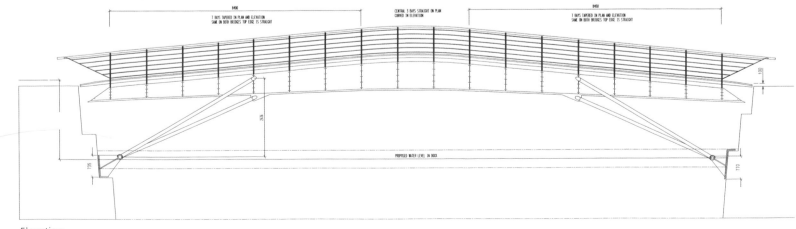

Elevation

CARDIFF BRIDGES

Both footbridges were conceived to span on legs that lightly touch the dock wall in such a way that, with the water maintained at a constant level by the barrage, they appear to stand on the surface of the water. The effect is similar to that of the water boatman, an insect so-named because of its ability to balance on the water surface.

The bridges span the listed Graving Docks in Cardiff Bay, and were constructed as part of the Cardiff Bay Development Corporation's regeneration plan for the area. The two spans vary, and a modular approach was developed to accommodate minor individual differences while at the same time enabling a standardised fabrication process. A third bridge was designed by Brookes Stacey Randall, but remains unbuilt.

The first bridge is clad with polished, stainless steel; the second has a PVC fabric mesh. The stainless steel provides a beautiful, glistening surface, responding to the reflected light from the water, while the fabric mesh permits a veiled view of the structure – at night the surface glows due to lights mounted within the body of the structure. This light is further supplemented with small diameter fibre-optic lights along each edge of the bridge.

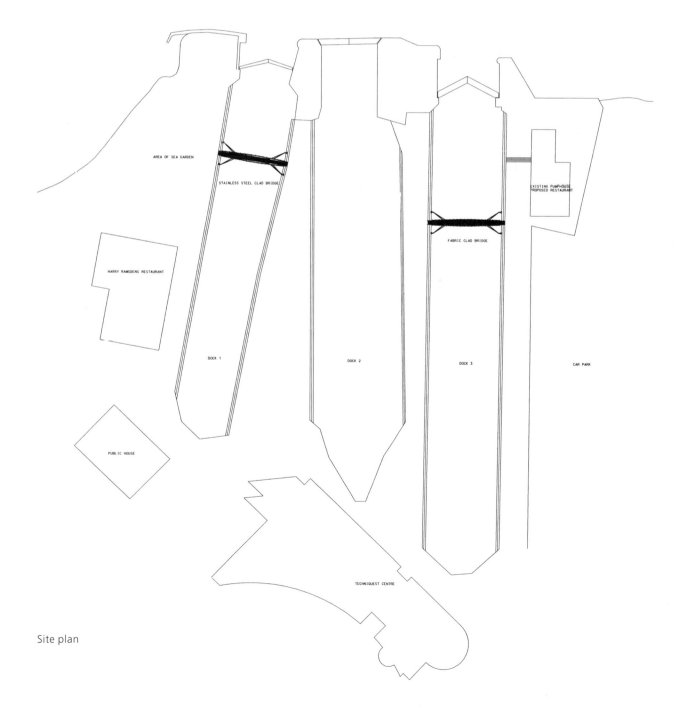

Site plan

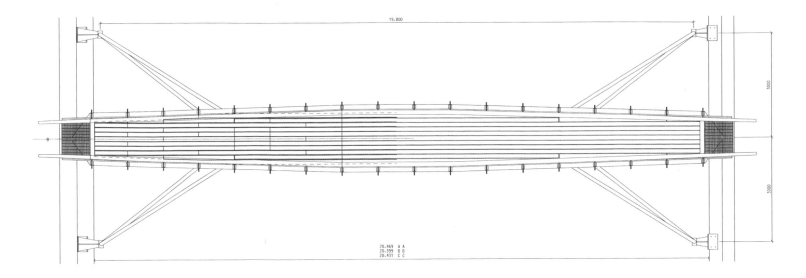

Plan

Model photo

THREE MILLS BRIDGE
London, England

Located in Bromley by Bow, this bridge is in a prominent position, facing a collection of eighteenth-century mill buildings to the south. The bridge spans 30m across the neck of the river, forming a visual closure to the river basin in front of the adjacent tidal mill.

It is a heavy-duty road bridge, with a pedestrian bridge slung along one edge. The principal spans are formed in welded plate girders; secondary steels span transversely between them. The road surface is constructed from open steel mesh and the pedestrian deck is made in a similar way but cantilevered from below the south girder.

Each girder is painted in a different colour scheme – the north girder is pale blue/grey, allowing its silhouette to blend with the sky on approach from the north, while the south girder is blue/black, reinforcing the closure of space towards the south. The tops of the girder flanges and the handrail are silver, gold, and copper in colour, giving three lines of reflection and light. The deck between is of transparent or pale grey steel, depending on the angle of view. The south girder has an apparently anarchic pattern of penetrations along the length of its welded plate, which undercut its solidity and admit sunlight and views from the pedestrian deck, through to the bridge deck and pale blue girder beyond. These penetrations are elaborated in an artwork that introduces shadow, colour, scale, and interlocking rhythms, marked by metal studs and sleeves.

Where the towpath on the east side is interrupted by the new bridge and road, a kite-shaped, granite-covered square has been created, focusing attention on the new entrance gates of a nearby sports ground. The road is allowed to traverse this square, yet the unified character and dynamic of the space gives predominance to the pedestrian. Diverging lines of granite slabs mixed with cobbles lead towards the central gate, which is characterised by increasing transparency towards the top.

The ensemble of bridge, square, and gates uses a common design language – made possible by a cross-project collaboration between the disciplines of engineering, architecture, and artwork.

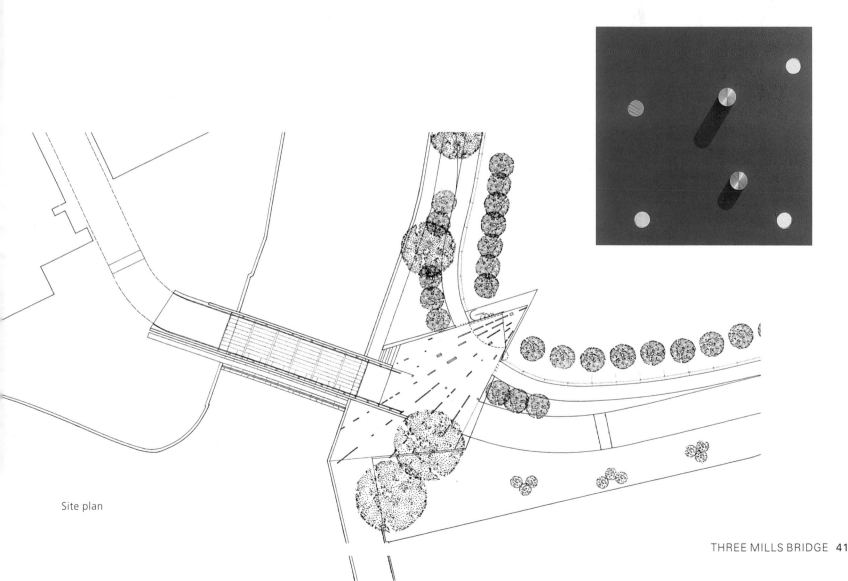

Site plan

Site sections

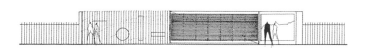

Gate, elevation and site section

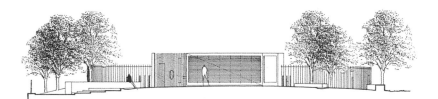

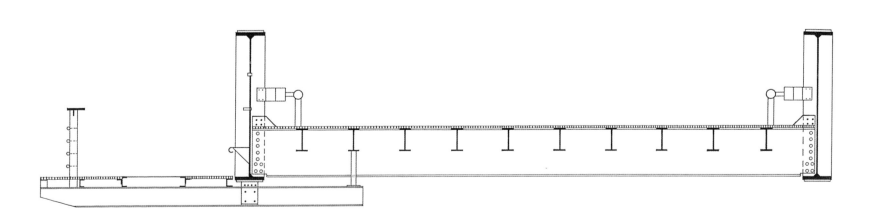

Section

CZWG
GREEN BRIDGE
London, England

Regeneration proposals for London's Mile End Park, which successfully obtained Millennium Commission Funding, included the idea of joining the two halves of the park with a landscape section. CZWG put forward the idea of the Green Bridge, which was adopted as part of the regeneration plan.

The bridge spans 30m and is 24m wide. Gradual inclines either side, at 1 in 20, lead up to some 7m above road level. These curving approaches respond to principal routes in the park, and mitigate the impact of some terraced buildings. The bridge abutments are occupied by shops and adjacent restaurant spaces, which enhance the vitality of the Mile End Road, emerging to the north-east and south-west as semicircular end abutments. The structure is laid out on a 7.2m wide grid (the industry standard retail width), with an optimum 2.5 times depth. It also carries the load of the bridge away from the tubeline tunnels under the site. The rental income of the shops is put back into lifelong maintenance of the whole park.

The underside of the bridge curves up conventionally across the span, but is also saddle-curved along the road. This has the effect of lifting the sides of the bridge to reduce the visual mass and appear less oppressive to those passing underneath.

For the bridge deck it was important to maintain the illusion of the park gliding seamlessly over the road, and the landscape here is as elsewhere in the park. Through this a walkway/cycleway continues across the bridge with grass verges either side. The cross-section of the bridge ensures that these verges are banked above the pathway, and surround the crossing to mask the road either side.

The landscape of the bridge and its slopes lead the eye onto fine landscape features to the north and south. A tumulus mound to the north has an inviting series of grass plateaus, while to the south wild flower terraces lead down to a fountain pool.

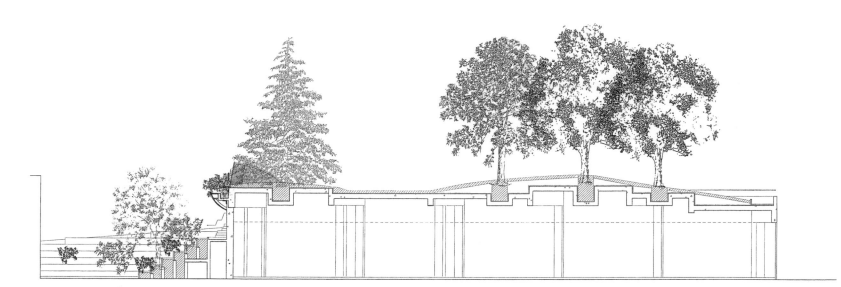

Section

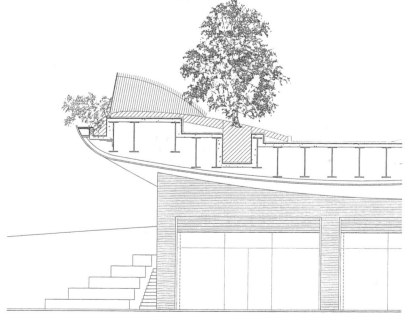

Cross section

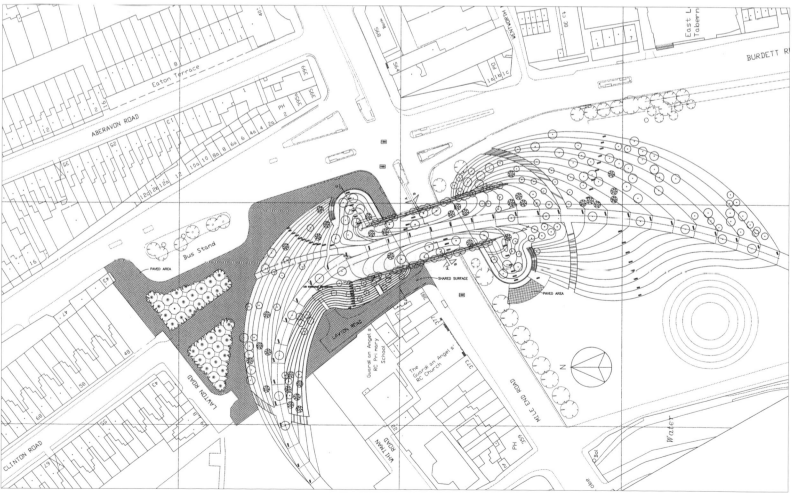

Site plan

TREE TOP WALK

Valley of the Giants, Walpole, Western Australia

Near the small rural community of Walpole, 430km south of Perth, is the Valley of the Giants – a site that has attracted visitors since the early years of the century, so-named because of the endangered Red Tingle trees that are unique to this isolated area. They reach heights of 50m and, at their base, the hollow trunks are up to 5m in diameter. The centrepiece of the redevelopment of this unique site is the Tree Top Walk and Tingle Shelter project. This award-winning project is the work of Donaldson + Warn Architects, in collaboration with structural engineers Ove Arup & Partners, and David Jones, an environmental artist.

Visitors arrive first at the Tingle Shelter, comprising two pavilions linked by a timber deck. These timber-framed buildings house the office, shop, and toilet facilities for the site. A timber-decked jetty leads from the shelter to the bridge trusses of the Tree Top Walk. This walkway, over six steel trusses supported by steel pylons, takes the visitor up into the tree canopy.

Eight pylons support 3m diameter platforms and the walkway bridge trusses that span between them. The trusses rise on an incline of 1 in 12, while the valley drops away below. At the highest point the visitor is 40m above the ground. The gentle slope ensures universal access to the tree canopy. The one-way walkway over the trusses is 900mm wide with 3m diameter platforms at the change of direction between each truss. The decking of the walkway is an open steel mesh that affords a view of the supporting truss below, and through that the forest floor. The circular platforms have steel floorplates to create the feeling of a more secure resting place. The balustrades and handrails are comparatively high to ensure safety, but are open to maintain the transparency of the whole structure.

Elevation

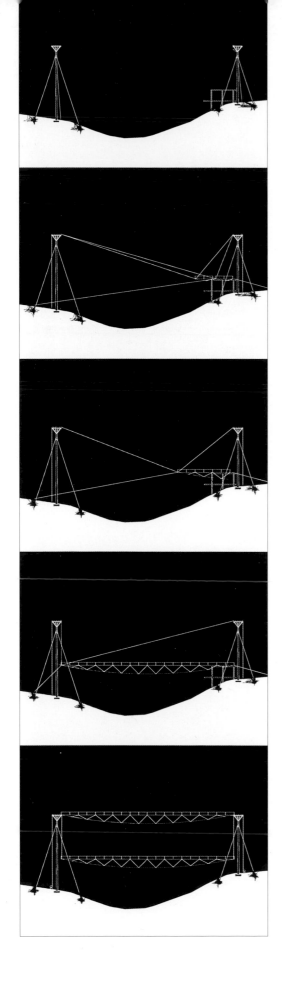

L-R Construction stages; plan

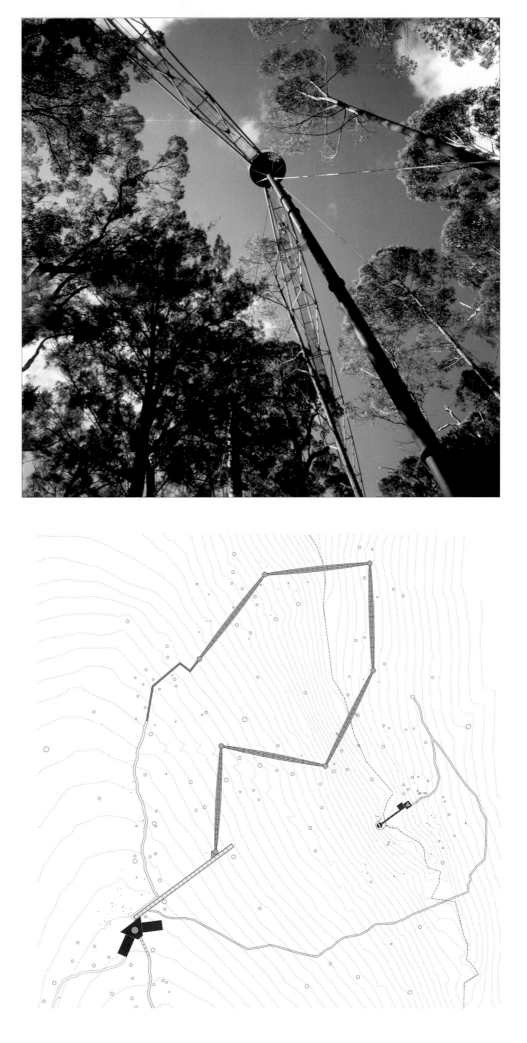

KOLISTE FOOTBRIDGE

Brno, Czech Republic

The bridge forms part of a 200m long pedestrian link that traverses a park, a large thoroughfare, and a new building to fill an empty space between the street frontage and an internal courtyard garden leading to the commercial complex of Unistav Centre, in Brno.

A clear span without any supports was required, commencing from the elevated platform in the park and crossing the road to the first floor of the new building. The need for the minimum vehicle headroom established the height of the new pedestrian link, and also the level of the passage through the building. Inside the park the new connection was to avoid protected trees and maintain the height that suited the existing entrance on the first floor to the Unistav commercial complex, and eventually to touch the ground of the existing garden; these requirements resulted in a curvilinear form.

The longest span is between the park and the new building. The structure was calculated to suit the maximum span and the supports in the park were positioned accordingly. The section is asymmetrical, with elevation of the main truss relatively high to minimise the sizes of individual structural members. The floor is formed by concrete panels with glass lenses, which are back-lit and give the walkway a distinctive glow during the hours of darkness. Illumination of the structural features also responds to the required lighting levels, and gives the bridge a special character.

The footbridge comprises four spans of a continuous truss, crossing the road, the courtyard, and passing through the new office development. The tubular steel structure is fully welded with gusseted joints and sliding bearings. The girder is formed as an asymmetrical torsion tube.

The bridge piers are tubular – steel, triangulated tripods, resisting vertical and lateral loads and the significant twists from the curving frame. On the west side of the bridge the span meets a steel platform, from which steel-framed stairs and a ramp lead down. On the east end the bridge enters the building's first floor level, bearing directly onto an existing building structure.

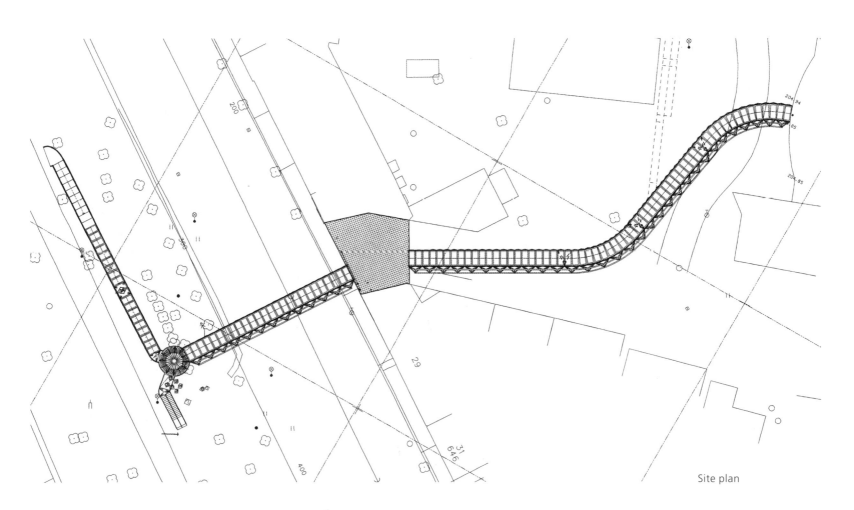

Site plan

MILLENNIUM BRIDGE

London, England

The Millennium Bridge springs from a creative collaboration between architecture, art, and engineering. Developed with sculptor Anthony Caro and engineers Ove Arup & Partners, the commission resulted from one of the most popular international competitions ever held. London's only bridge solely for pedestrians, it is the first new Thames crossing since Tower Bridge in 1894.

The bridge links the City and St Paul's Cathedral, to the north, with the Globe Theatre and Tate Modern on Bankside to the south. A key element in London's pedestrian infrastructure, it has a social and economic impact on both sides of the river, creating new routes into Southwark and contributing to its regeneration, as well as encouraging new life on the embankment alongside St Paul's.

The paramount concern in designing the bridge was the experience that it offered the user. A windscreen shelters those crossing, allowing them to pause comfortably and enjoy London from a new vantage point, free from traffic and fumes. The materials selected – stainless-steel balustrades and aluminium decking – will be enhanced by use and wear.

Structurally, it is a statement of technological capabilities at the beginning of the millennium. Spanning 320m, it is a very shallow suspension bridge. Two Y-shaped armatures support eight cables which run along the sides of the 4m wide deck. Steel transverse arms clamp onto the cables at 8m intervals to support the deck itself. This groundbreaking structure means that the cables never rise more than 2.3m above the deck – a distance ten times less than that of any other suspension bridge – allowing pedestrians uninterrupted panoramic views of London and preserving sightlines from the surrounding buildings. As a result, the bridge has a uniquely thin profile, forming a slender arc across the water. It spans the greatest possible distance with the minimum means. By day it appears as a thin ribbon of steel and aluminium; at night it forms a glowing blade of light.

The bridge as a minimal intervention an elegant blade - steps & ramps connect to the banks - to walk thru/ under/over/around - platforms over the water to view & browse

The axis of the chimney. The symbol of the power station & the New Tate - an existing marker

the New globe

The axis of the New crossing a New marker & a New link - North/South Art/commerce

Bankside & the bridge - symbols of regeneration.

The "place" of the New Tate - outside

The axis of the "light box" - the symbol of the New Tate - a New marker.

Conceptual drawing

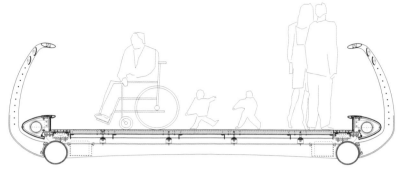

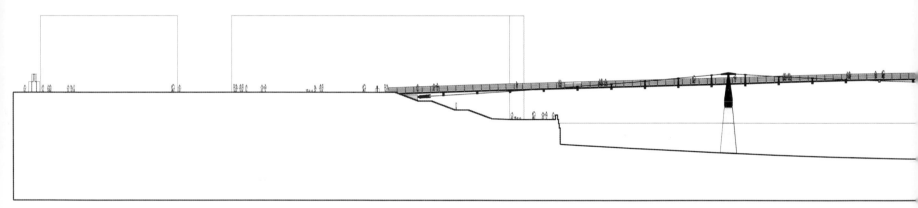

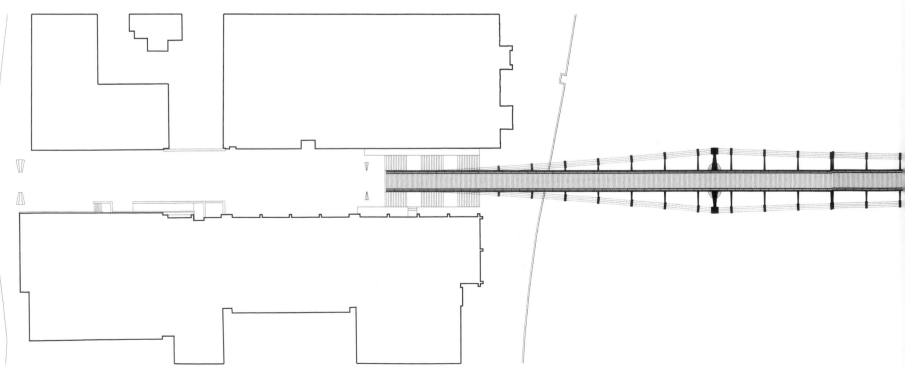

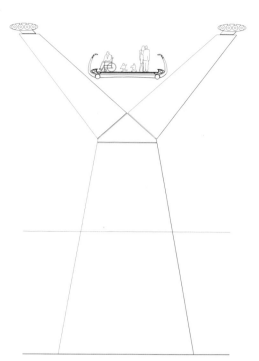

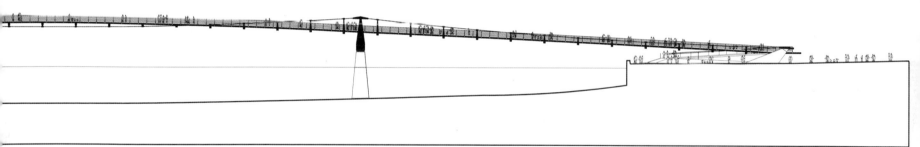

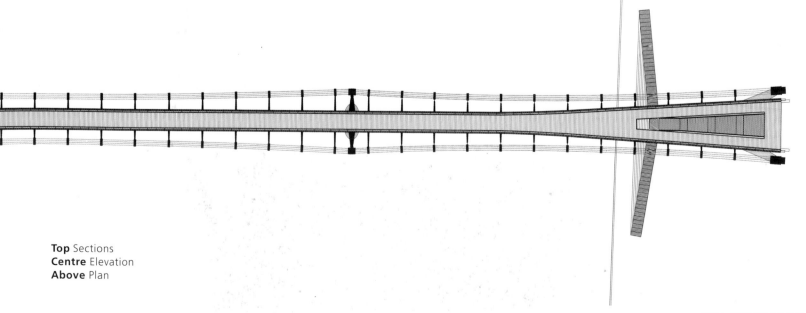

Top Sections
Centre Elevation
Above Plan

DOGANA GATE

Republic of San Marino

Marking the boundary between one territory and another, the Dogana Gate (Porta di Dogana) alerts travellers to cultural and political differences, celebrating the act of entry and reacting against the pernicious present-day tendency to make all places the same. The gate is located on one side of the main highway to Rimini, the principal link between the Republic of San Marino and Italy.

The gate takes the form of a double pole structure raised on one side of the road, supported by both a vertical and inclined stay. While the vertical stay is in effect a simple guy, the inclined stay is made up of several interconnected elements. Of these the main form is hinged from each of the two poles and suspended by a guy – the boom is then anchored to the ground on the other side. The architrave of the gate is supported by a spar 30metres tall, bearing above it the flag of the republic being left behind,

the flag of the republic being entered, and the flags of the nearby castles. It stands proud against the sky, elegantly proclaiming the boundary between the two territories.

Not only is it impressive by virtue of its size, there are also other, more subtle elements that link the gate to the collective imagery of the citizens of San Marino. There is an evident analogy between its form and that of a ship, recalling the coast and sea which the Republic never reached but which its citizens continue to observe. This appears in the metal structure, the stays, the steps running all the way to the top, and the two broad sails flanking the catwalk across the Superstrada. The gate also evokes the image of trees, their branches stretching naturally upward and outward, as can be seen in the wooded areas of Montefeltro.

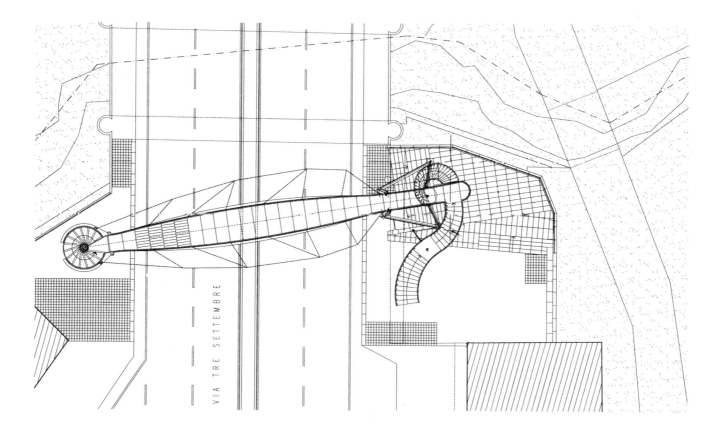

Site plan

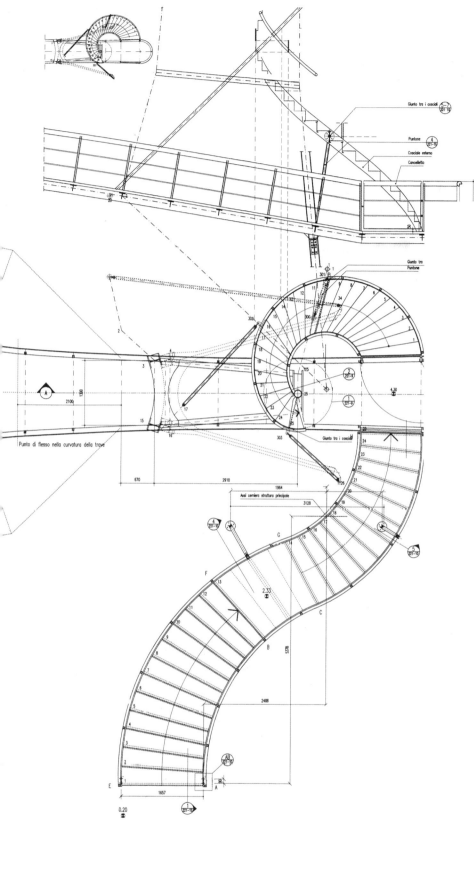

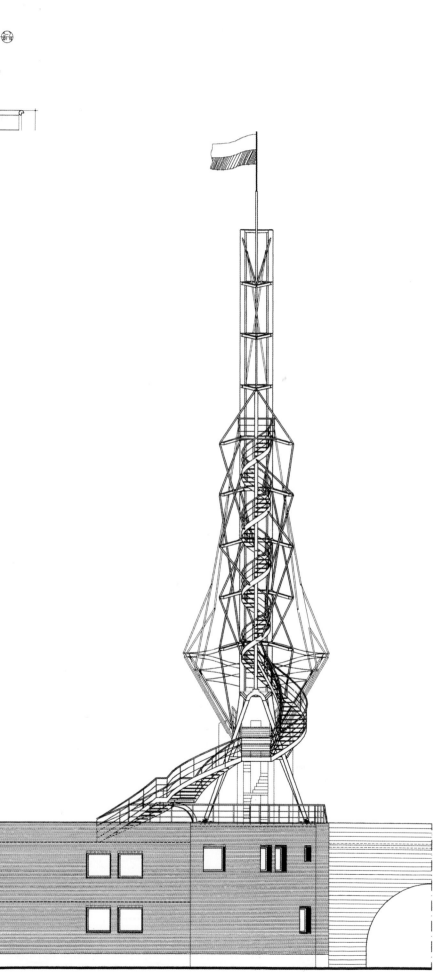

Above Plan of staircase
Right Section

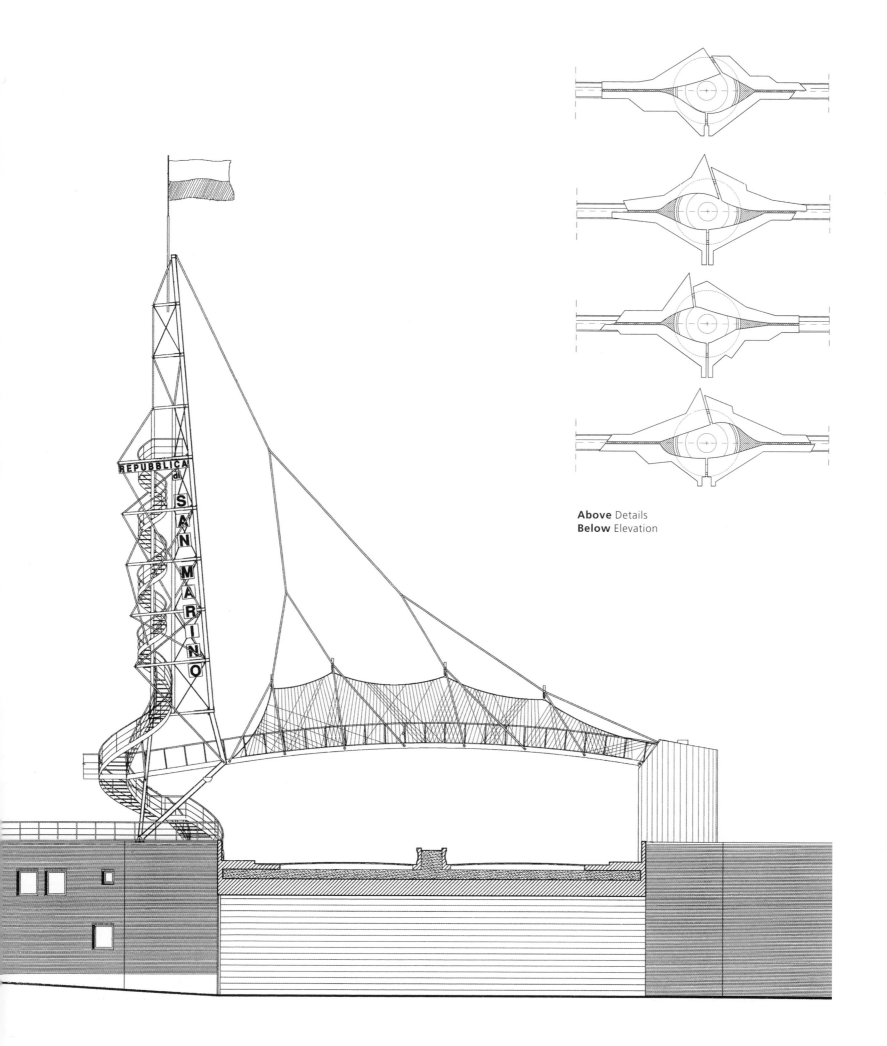

Above Details
Below Elevation

REPUBBLICA di SAN MARINO

FOOTBRIDGE OVER THE RIVER MUR

Graz, Austria

The design for a footbridge over the River Mur arose from an invited competition, organised for teams of architects and engineers by the municipal council of Graz.

The competition focused on the urban significance of the project, prompted by the need to connect the old part of the town with the former Mur suburb. This link was intended to heighten activities near the river, and contribute to the historic townscape context.

For this winning design, the architects used a fine glass balustrade and light construction system so that views along the river were not disrupted by structural elements. The bridge comprises a single-span beam, triangular in cross-section – this geometrical approach was employed to overcome the height difference between the banks. The steel structure emphasises the appropriateness of the bridge as a lightweight intervention in the historical context of its surroundings.

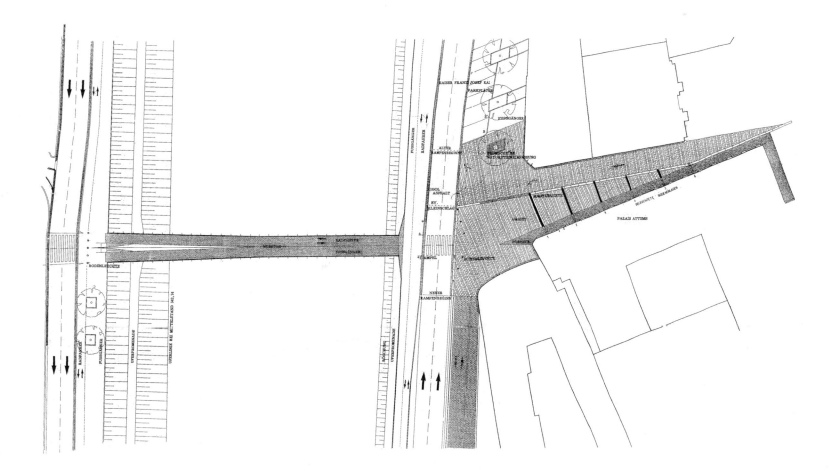

Site plan

CORPORATION STREET FOOTBRIDGE

Manchester, England

On 15 June 1996 a large bomb exploded in Manchester, injuring 220 people and causing immense physical damage to the core of the city centre, its social and economic fabric. The shattered footbridge, which connected two shopping centres across Corporation Street, remains one of the most vivid images.

The Corporation Street Footbridge represents one of the world's first bridges whose delicate structure traces a hyperbolic paraboloid of revolution. It spans 19m, with 18 rods pre-stressed against 18 compression members via rings at each end. The enclosing glass technology is equally revolutionary, utilising a polyethylene terethalate interlayer clamped into elliptical castings. The footbridge represents a particular response to a unique civic context, and has quickly become a potent iconic symbol for the regeneration of Manchester's city centre after the bomb.

Corporation Street is a canyon-like road and is a significant, linear, north-south route through the city, culminating in the civic space of Albert Square. The footbridge appears as a lightweight glazed membrane stretched across the street, presenting a minimal intervention with the surrounding urban context. Its transparency is heightened by the arch which permits uninterrupted views, and whose symmetry optically redresses the change in level of the boardwalk. The expressed steel structure is intended to impart the grain of city scale externally, whereas the membrane is a smooth, tactile, finer grain for people using the bridge.

The deck or boardwalk is a conventional steel structure which spans across the varying width, and the bridge is finished in American oak. The void below allows air to be admitted via the compression rings at each end: heated in winter, the warm air rising at the edges of the boardwalk and venting once again through the compression rings at high level. Ventilation in summer is by similar means, and thus the profile of the bridge also optimises the pattern of natural ventilation. The glazing system, designed in conjunction with Arup Façade Engineering, comprises purpose-made elliptical stainless-steel castings, each clamping six triangular sheets of laminated glass panels.

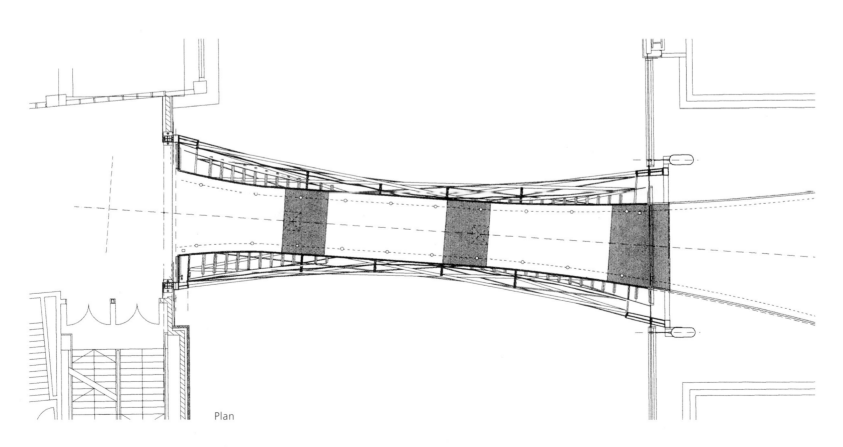

Plan

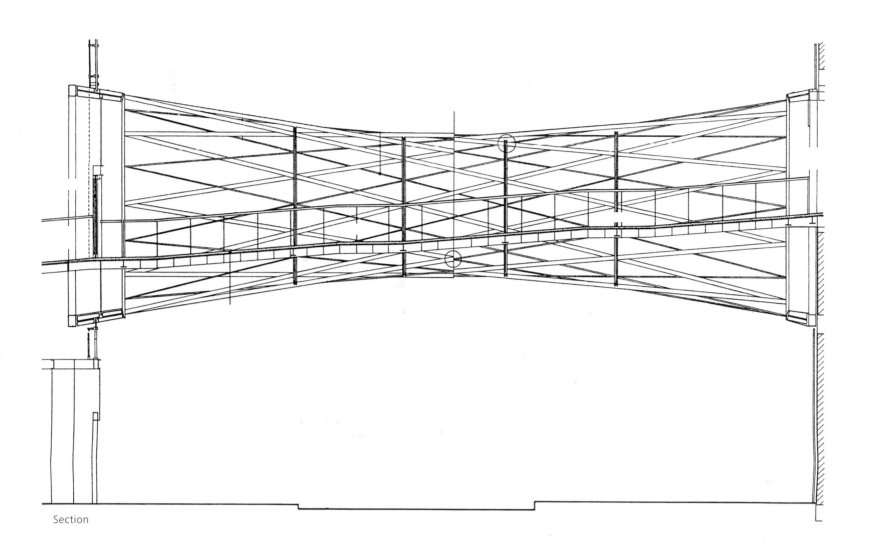

Section

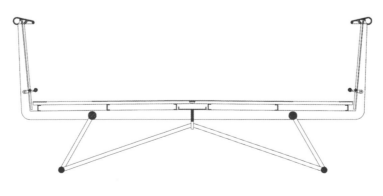

Section

LIFFEY PEDESTRIAN BRIDGE

Dublin, Republic of Ireland

Dublin's new pedestrian bridge, known as the Millennium Bridge, was the outcome of an international competition. The winning entry, out of 157, was designed by the Dublin-based Howley Harrington Architects, in association with Price & Myers Structural Engineers.

The bridge is a dynamic, contemporary design, being lightweight, transparent, and structurally daring, while fitting sensitively into its prominent urban setting of Dublin's historic quays. At the point of crossing the river is 51m wide, although the central span of the bridge is 41m between the projecting abutments. The truss is designed as an asymmetrical parabolic arch, the booms of which are solid steel rods, curving gently inwards from the abutments on each bank.

The rounded, stone-clad, concrete shells of the abutments act as spreadwaters to allow the smooth passage of water around them at all tide levels. From the quay wall, the pavement sweeps out over the river, opening up the bridge in width at each end. This relieves pressure at the busiest pedestrian points, providing a place to gather and wait before crossing the adjoining roads. With a gentle gradient of 1 in 20, the bridge allows those with wheelchairs and children's buggies to get across easily.

Terminating in the large concrete haunches contained within the granite-clad shell abutments, the bridge structure is a simple, but very efficient, portal frame. This permits the use of slender structural members, creating a sense of lightness and transparency, echoed in the balustrade of the deck and in the abutment railings. The slotted aluminium deck is supported by a series of secondary ribs running between the cross members. These are integral with the top booms of the truss, and continue upwards to provide supports for the balustrade and aluminium bronze leaning rail.

To allow safe passage across the bridge at night, the aluminium deck is lit using hidden emitters, supplied by fibre optic harnesses running inside the specially shaped bronze handrails. The abutment pavements are flooded by reflected light from the underside of the specially designed stainless-steel lamp posts, mounted on reconstituted granite bollards. Along the centre of the deck are a series of small uplighters, similar to runway lights. Finally, narrow beam floodlights are mounted under the abutment pavements to highlight the aesthetic qualities of the graceful truss.

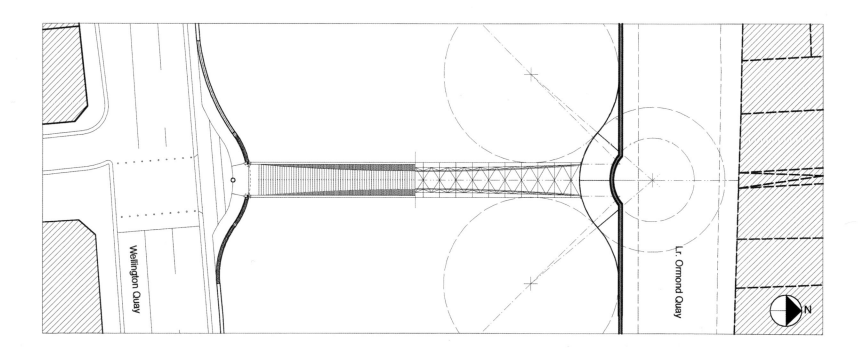

Plan

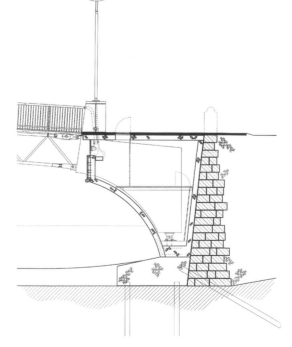

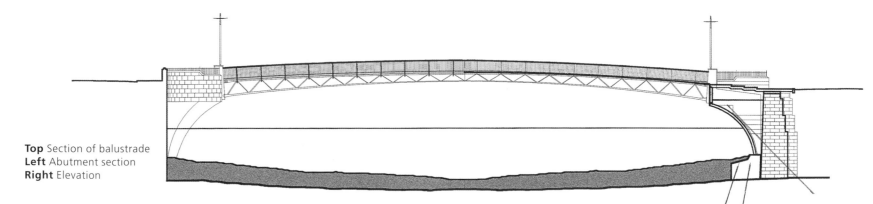

Top Section of balustrade
Left Abutment section
Right Elevation

Concept sketch

NATURAL HISTORY MUSEUM ECOLOGY GALLERIES

London, England

Above the spectacular back-lit vista of the entrance to the new Ecology Galleries, bridges leap across the asymmetrical space. Crossing the bridges people appear and disappear through two glass walls, one curved, one straight, designed to evoke the tendency of things to be continually dynamic while striving for equilibrium.

As visitors move down the vista, they are subtly drawn into the context of the exhibition by means of a layered, curved glass wall on the right, lit from behind by cool-coloured temperature lamps that suggest the sense of a glacier (representing water). To the left, the straight glass wall is also back-lit in a similar manner, but utilises warm-coloured temperature lamps to convey the energy of fire. The visitor's attention is drawn to the first bridge above, which is made from treated glass set onto an organic structural form, and registers this fragile element and the image of people passing over.

The visitor then passes beneath three further bridges, which represent the evolution of man's manipulation of the planet's natural resource. The first bridge presents a spectacular view back towards the entrance area, and has an 'earth' floor made of recycled rubber. Continuing through the exhibition spaces the visitor crosses the second bridge, whose surface is wood, and again can enjoy the central space of the gallery. The surface of the third bridge is metal. The visitor then crosses the central space for the last time, across a surface of glass decorated with Ginkgo biloba and chestnut leaves. (Ginkgo leaves are the sole survivor of a genus hundreds of millions of years old, while chestnut leaves are a complex leaf form of much more recent times.) The bridges are united by a cherry handrail, which continues throughout the exhibition route.

All three bridges are supported by a pair of braced hollow tubes, 165mm in diameter. These sleeve through the rib-like fins of the walkway supports, whose ends turn up like antlers to form the posts for the handrail. Two stainless-steel 30mm cables run under the structure, acting as external pre-stressing tendons.

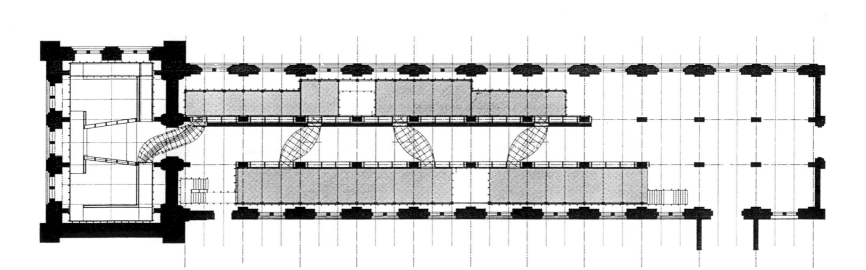

Plan

Below Axonometric of main elements
of exhibition space and floors
Below left Concept sketch

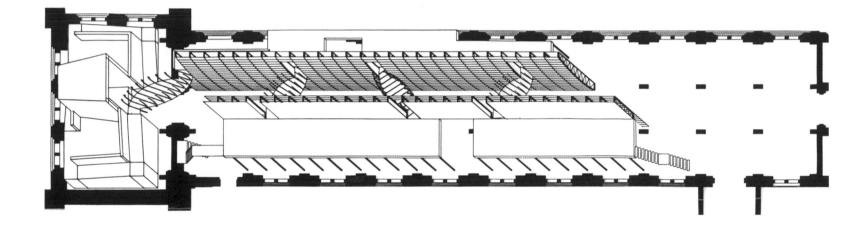

JIRI STRASKY

ROGUE RIVER PEDESTRIAN BRIDGE

Grants Pass, USA

This bridge, which opened in spring 2000, is located in a popular recreational area, and connects a city fairground with a sports field across the scenic Rogue River, Grants Pass, Oregon. The bridge crosses not only the bed of the river, but also a protected wetland area on the right bank. It also carries both a sewage main and a waterpipe.

The Rogue River Pedestrian Bridge consists of a stress-ribbon concrete deck of three differing spans, 73.15m, 84.73m, and 42.67m respectively, supported at the abutments and on two piers. The concrete deck forms a catenary shape between the pier supports, the slope of which is variable, with a slope of up to eight per cent and maximum sag of 1.54m. The width between the railings is 4.27m. In the middle of those spans above the river and over the wetlands are situated observation platforms to encourage recreational use of the bridge.

The deck is a composite of pre-cast concrete elements, with a cast-in-place concrete portion. The pre-cast concrete deck elements consist of waffle-slabs, 4.70m wide and 3.05m long. The pre-cast concrete elements are designed to be suspended on bearing cables, situated in the areas to be filled by the cast-in-place concrete. The prestressing cables, post-tensioned after the casting of the deck slab, give the structure its required stiffness. The large horizontal forces in the stress-ribbon deck are resisted by a combination of pin-piles and tie-backs. The bridge is surrounded by trees, with a heavily wooded area existing on both banks. Within the bed of the river itself, the piers correspond to the branching of trees.

The bridge demonstrates an innovative construction process. Firstly, anchor blocks and piers were erected, followed by pier segments. The bearing cables were then placed from abutment to abutment and across the two piers. The pre-cast segments, along with the prestressed cables and reinforced steel, were then lifted into position along the entire length of the bearing cables. Using the pre-cast segments as a formwork, the deck was cast in one operation – and thus the deck is continuous from one abutment to the other without any expansion joints.

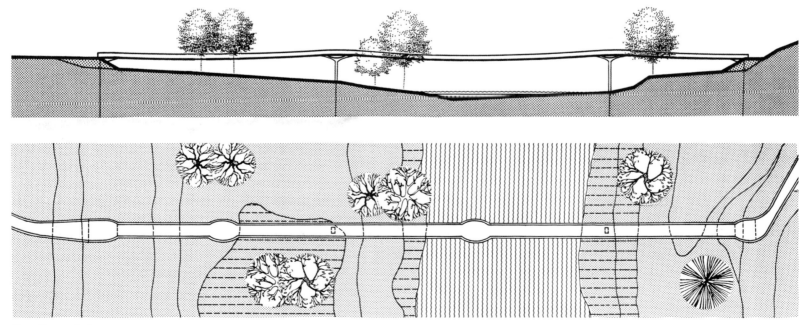

Elevation and plan

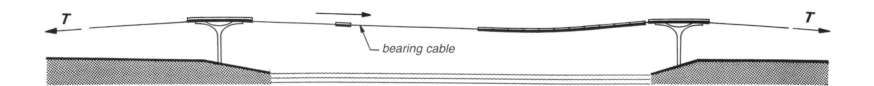

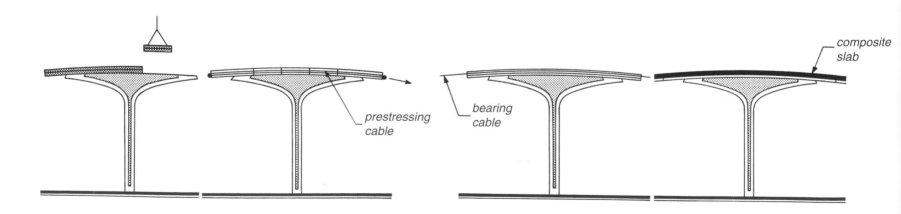

bearing cable

composite slab

prestressing cable

bearing cable

Construction diagrams

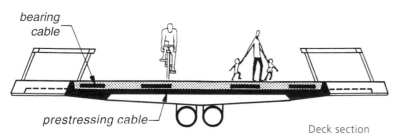

bearing
cable

prestressing cable

Deck section

observation
platform

View of deck
from below

JIRI STRASKY
SACRAMENTO RIVER TRAIL PEDESTRIAN BRIDGE
Redding, USA

The Sacramento River Trail and its connecting bridge form part of the park system of Redding, California. This area of the park lies to the northwest of the city on both sides of the river, extending 4km upstream to the Keswick Dam. The new bridge provides a link between previously separated trails lying just above the rocky areas on each side of the river. Because of the dam's presence, the riverbanks directly downstream have extensive rock outcropping which adds dramatically to the beauty of the basin. To preserve this natural terrain, and to mitigate adverse hydraulic conditions, it was important to avoid founding any piers in the river basin.

The bridge is formed by a ribbon of prestressed concrete over a span of 127.40m and fixed at both end abutments. The deck width between railings is 3.42m, while the total width of the structure is 3.96m. The sag at mid-span goes up to 3.35m. By keeping the abutments at the same elevation with a minimal drape in the centre, the slope at the ends is held

to an acceptable nine per cent. Considerable prestressing material in the superstructure (236 strands of 13mm diameter), plus large rock anchors embedded deeply into the hillside at both ends, were required to form this shallow drape. Bridge vibration studies were carefully considered in the design for a wide range of frequencies, including those generated by jogging and the remote possibility of vandals attempting to excite the bridge physically.

Construction of the superstructure consisted of lifting the pre-cast segments onto the bearing cables and sliding them into their final position. Placement of additional cables directly over the bearing cables, in two troughs, casting them in place and further stressing provided the required stiffness for the bridge. A light bridge of a stressed-ribbon design with a single span and no joints presented an elegant solution, facilitating a simple erection without the need for construction in the river basin.

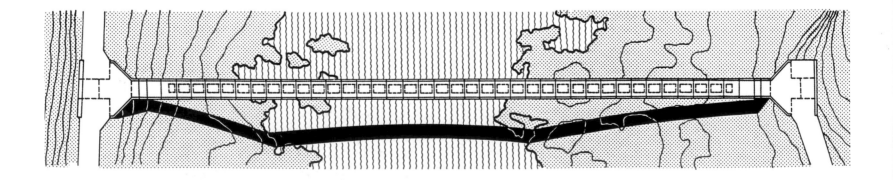

Plan

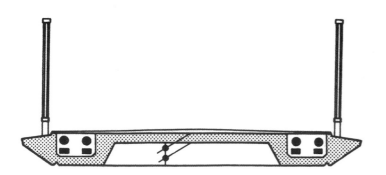

Cross section

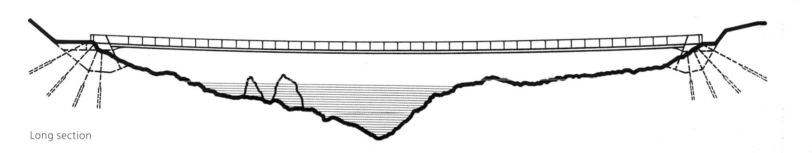

Long section

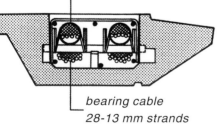

prestressing cable
31-13 mm strands

bearing cable
28-13 mm strands

Cross section detail

VRANOV LAKE PEDESTRIAN BRIDGE
Czech Republic

The pedestrian bridge over the Swiss Bay is located in the recreation area of Lake Vranov, Czech Republic. The bridge connects the public beach on one side with restaurants and shops on the other, and serves as a carrier of water and gas pipelines. The bridge was designed as a one-span structure with a horizontal clearance of 252m, a suspension structure fitting best with the setting and proving to be economical to construct. The arched deck (a mere 40cm in depth) is suspended on two inclined suspension cables of three spans, respectively of 30m, 252m and 30metres. The cables are deviated in steel saddles, situated at the diaphragms of A-shaped concrete pylons and secured in anchor blocks.

The deck is assembled from pre-cast segments of double 'T' cross section, and is post-tensioned in the supports that are linked by pre-stressing rods to the anchor blocks. The two main cables consist of 108 strands, 15.5mm in diameter, grouted in steel tubes. The hangers are steel rods of 30mm diameter, pin-connected with the deck and suspension cables.

The completed bridge shows how an efficient combination of pre-cast concrete and post-tensioning technology can produce a construction that is extraordinary slender, beautiful, and at the same time safe and comfortable for pedestrians. The deck, with its span to depth ratio of 630m, rates among the most slender bridges ever built.

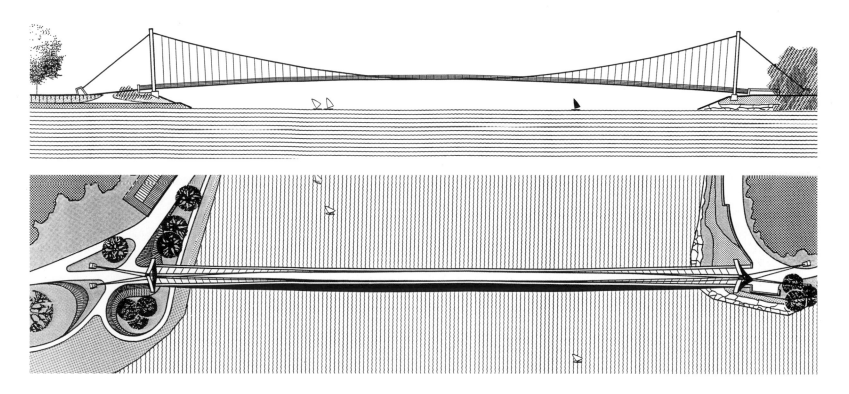

Elevation and site plan

Computer model

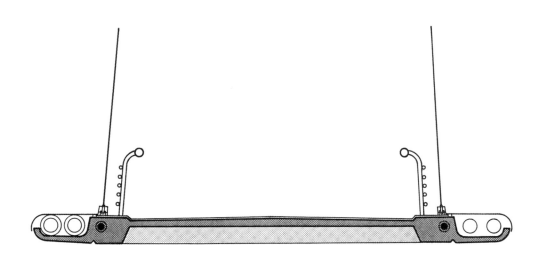

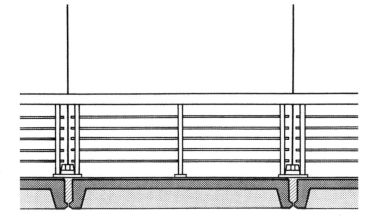

Deck section and elevation

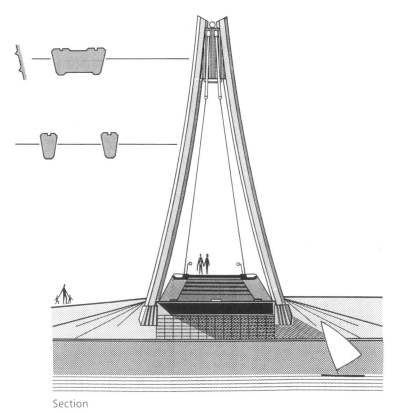

Section

Computer model

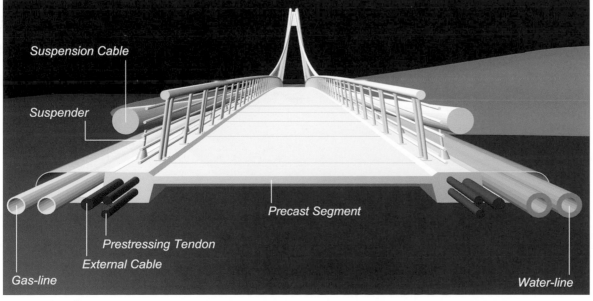

Suspension Cable

Suspender

Precast Segment

Prestressing Tendon

External Cable

Gas-line

Water-line

Construction diagram

JIRI STRASKY

WILLAMETTE RIVER PEDESTRIAN BRIDGE

Eugene, Oregon

Opened in spring 1999, this suspension bridge across the Willamette River is formed of a very slender concrete deck, flexibly anchored at abutments and hung on suspension cables formed by mono-strands grouted in steel tubes.

The overall length of the bridge is 178.8m consisting of two parts – pre-cast spans and *in situ* approach spans, the two elements generating in effect a platform and stair. The suspended spans are 23m and 103m for side spans and main span respectively. The entire bridge consists of five spans, including a curved ramp into a park on the east end. The average deck width is 6.5m.

The typical deck section of suspended spans consists of pre-cast concrete segments, 3m long and longitudinally post-tensioned together after erection. Two edge girders and the deck slab form the cross-section of the segments, with transverse diaphragms stiffening the segments at the joints.

The design of the bridge necessitated the inclusion of an observation platform in the middle of the central span. Therefore the main cables are threaded through the central wider deck panels of the main span.

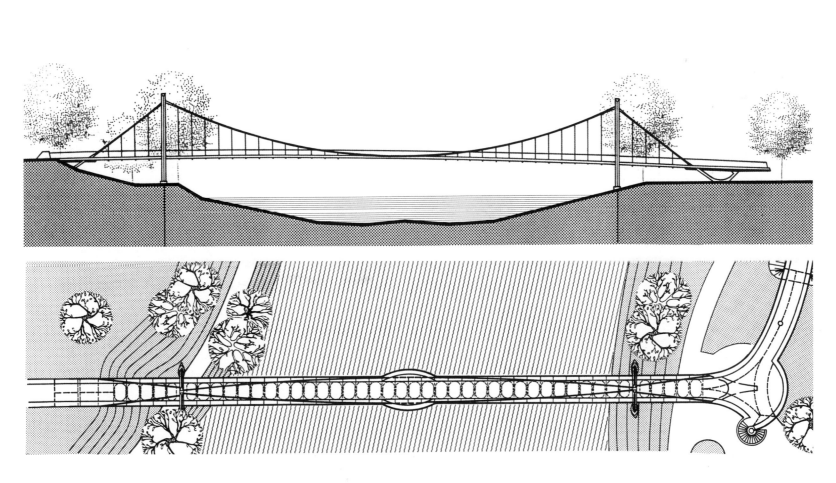

Elevation and site plan

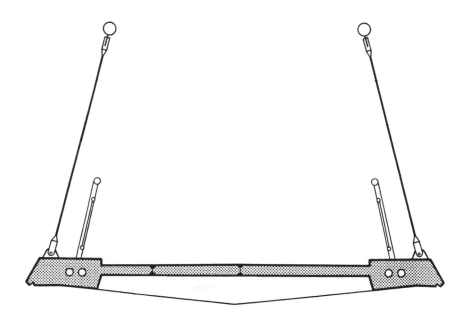

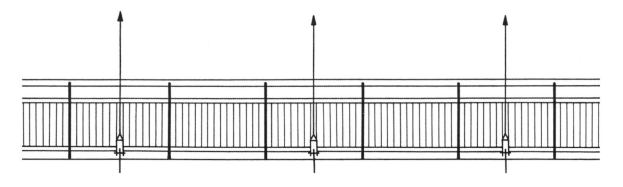

Top Detail section
Above Detail elevation

LINKBRIDGE 2000

London, England

Part of the Thames Riverwalk and Route One of the National Cycle Network, this illuminated sculptural footbridge fulfills various functions; firstly as a footbridge, designed to provide pedestrian access over a floodwall that obstructs the Thames Riverwalk. Secondly, it has a cantilevered platform which gives spectacular views over the Thames, including the Thames Barrier, Woolwich Ferry, and the dramatic Tate & Lyle factory. Thirdly it has become a landmark sculpture which can be seen from both sides of the river.

The main structure and balusters are fabricated in white powder-coated mild steel, with galvanised handrails and steps. The primary form creates a spine under the rising steps on either side, meeting a circular ring under the central platform. A third spine or leg descends at the front. The three spines act as a tripod supporting the cantilevered steps and platform. The balusters meet a steel mast, which cantilevers a lighting reflector. This bounces light on to the bridge. The bottom steps were manufactured in reconstituted Portland Stone.

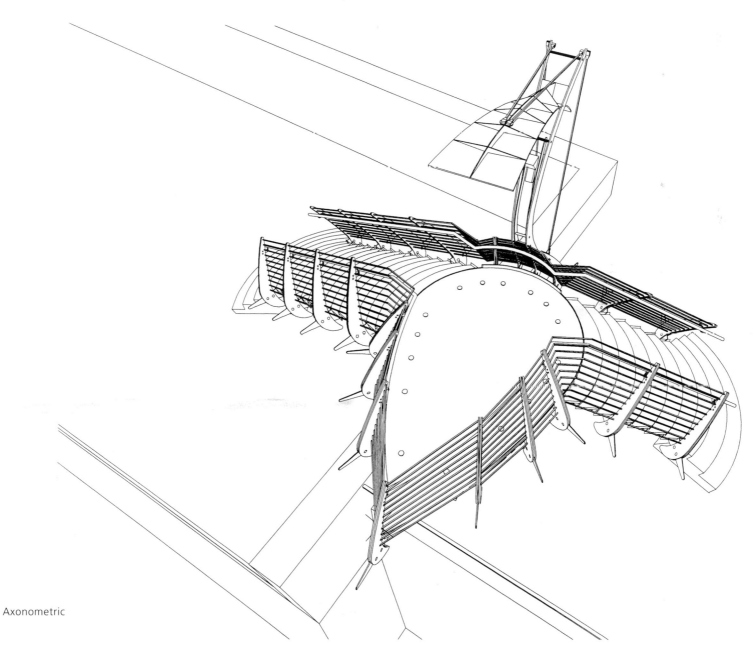

Axonometric

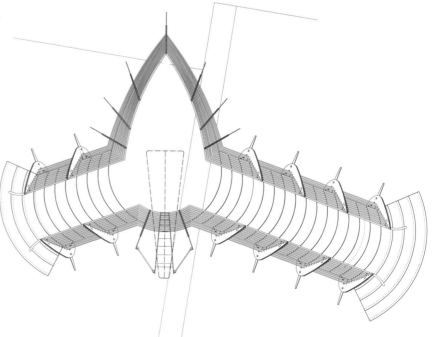

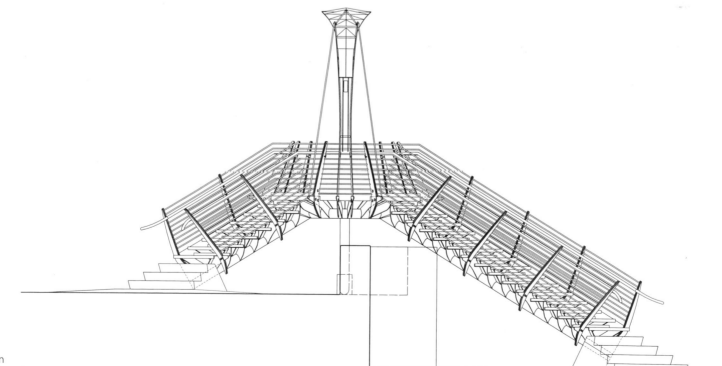

Left Plan
Right Elevation

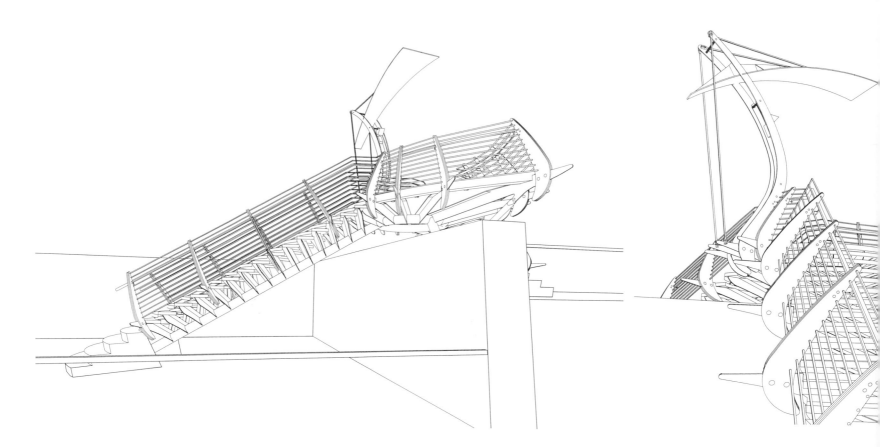

Perspectives

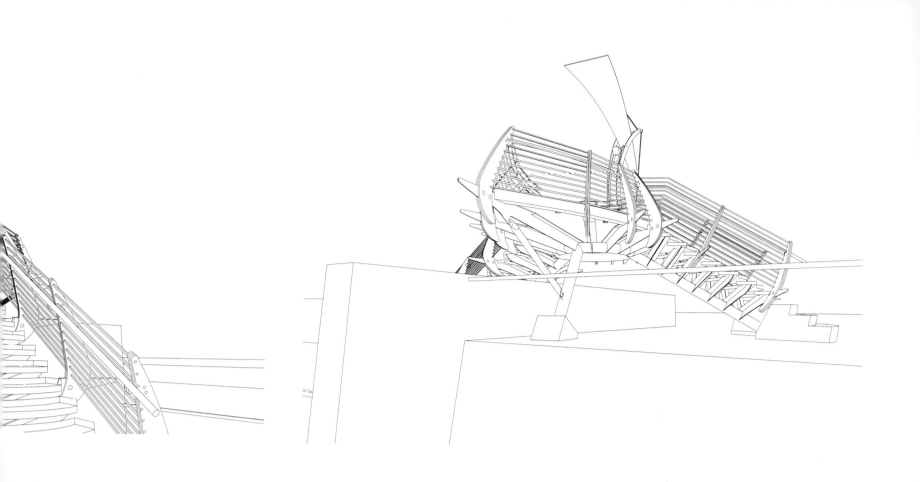

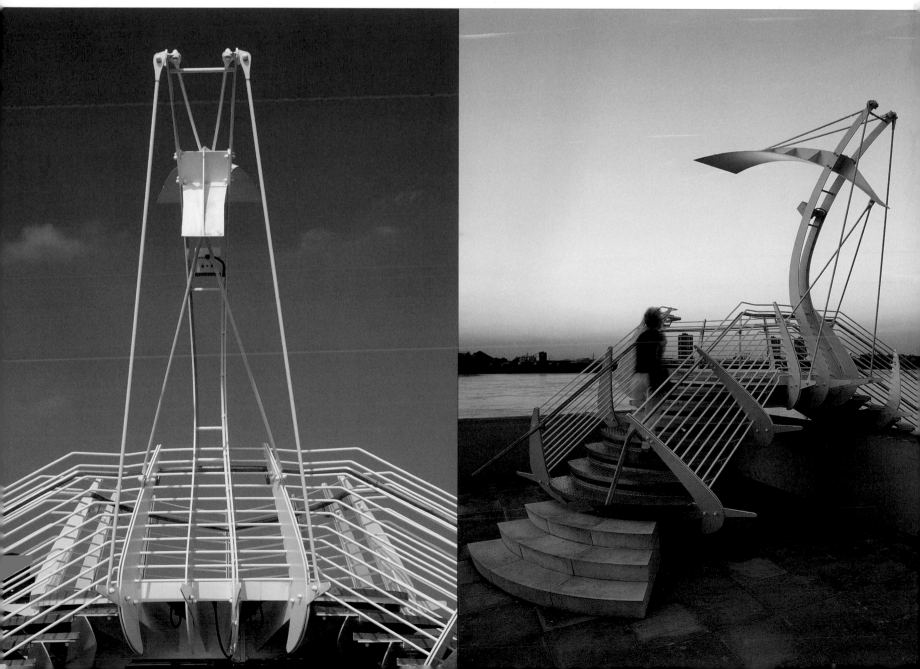

LIFSCHUTZ DAVIDSON
HUNGERFORD BRIDGE MILLENNIUM PROJECT
London, England

This high-profile project which won an international competition in 1996, involves the construction of two new pedestrian footbridges across the River Thames. Located on either side of the existing Charing Cross Rail Bridge, the new structures link the Victoria Embankment with the South Bank and Waterloo.

The design approach seeks to make best use of what exists on site, and the final form acts as a foil to the heavy railway structure behind. The use of inclined pylons pays homage to similar structures created for the 1951 Festival of Britain held on the South Bank, but takes advantage of developments in structural analysis to create an elegant lightweight structure. The concrete bridge decks are suspended over the river on an array of supporting steel bars, which vary in size to reflect the load paths. The physical connection to the South Bank Centre has been improved by linking the main downstream bridge termination stair and lift to the Royal Festival Hall terrace.

The materials used for the bridges have strength and durability. Steel is used for the pylons and cables, with the bridge decks formed in reinforced concrete. Each deck was cast on the South Bank and launched incrementally in 500m lengths. The deck is finished with a silver-grey wearing course, and the balustrades are in satin polished stainless steel.

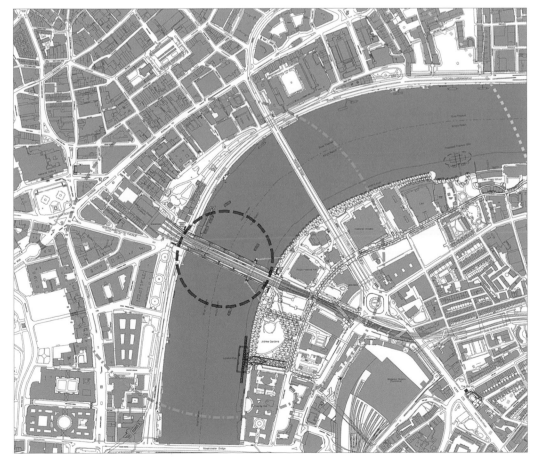

Location plan

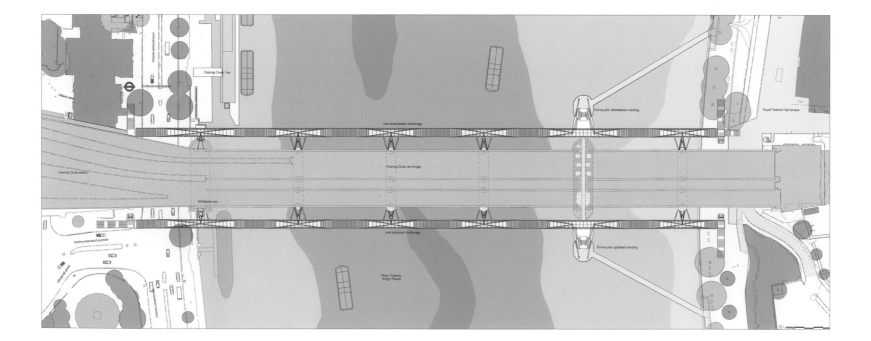

Plan

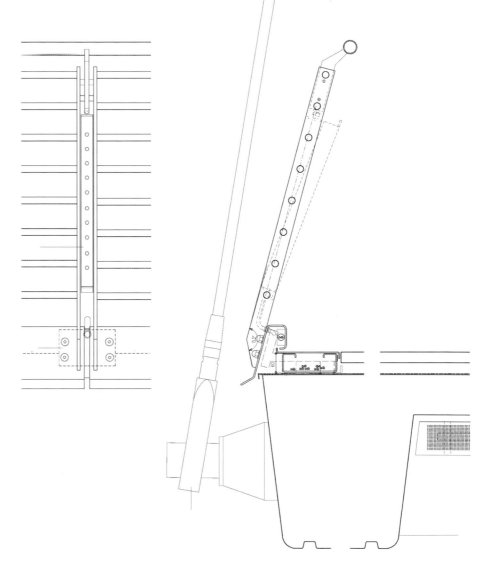

Top Section
Right Detail of balustrade

ROYAL VICTORIA DOCK BRIDGE PROJECT

London, England

In 1995 London Docklands Development Corporation initiated a competition for a new footbridge to cross the Royal Victoria Dock in East London. Architects Lifschutz Davidson and engineers Techniker won the competition, with proposals for an elegant cable-stayed transporter bridge incorporating a high-level walkway.

The visual appearance of the bridge reflects nautical and industrial themes of the dock's recent past, while also hinting at its future use for leisure and sailing, with the aid of marine elements including masts, cables, and hardwood decking. The dock has been designated as a national sailing centre, and to accommodate this the bridge deck is raised approximately 15m above water level and provides a single clear span of 130m.

The structure has a fully prestressed steel frame and was erected with minimum site bolting. Prefabricated sections, sized to the limits of road transportation, are connected across giant spigot pins and pulled together by external tendons. The structural form is based on a cable-stayed Fink truss with plate girder panel beams. In this case the truss is inverted and presents a radical departure from conventional cable-stayed fan or harp arrangements. The deck panels are fabricated from steel plate welded into a monocoque structure. The design uses high strength materials to accentuate the lightness of the design.

The bridge finishes have been selected for low maintenance combined with richness and quality, and include hardwood decking and perforated stainless-steel clad parapets and stair towers. The bridge is illuminated to complement the structural forms with mast top projectors highlighting pylons and cables. The ribs of the bridge deck soffit and balustrade are accentuated by concealed strip lighting, and the stair cores glow with blue light, which is reflected in the dark waters of the dock.

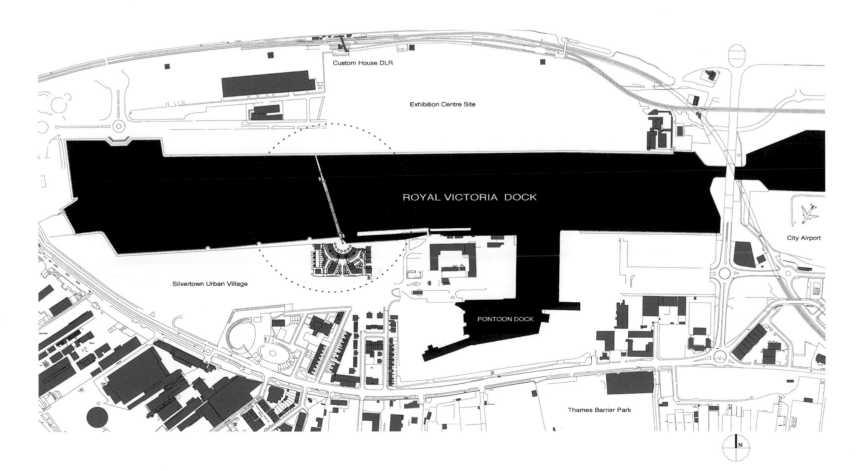

Site plan

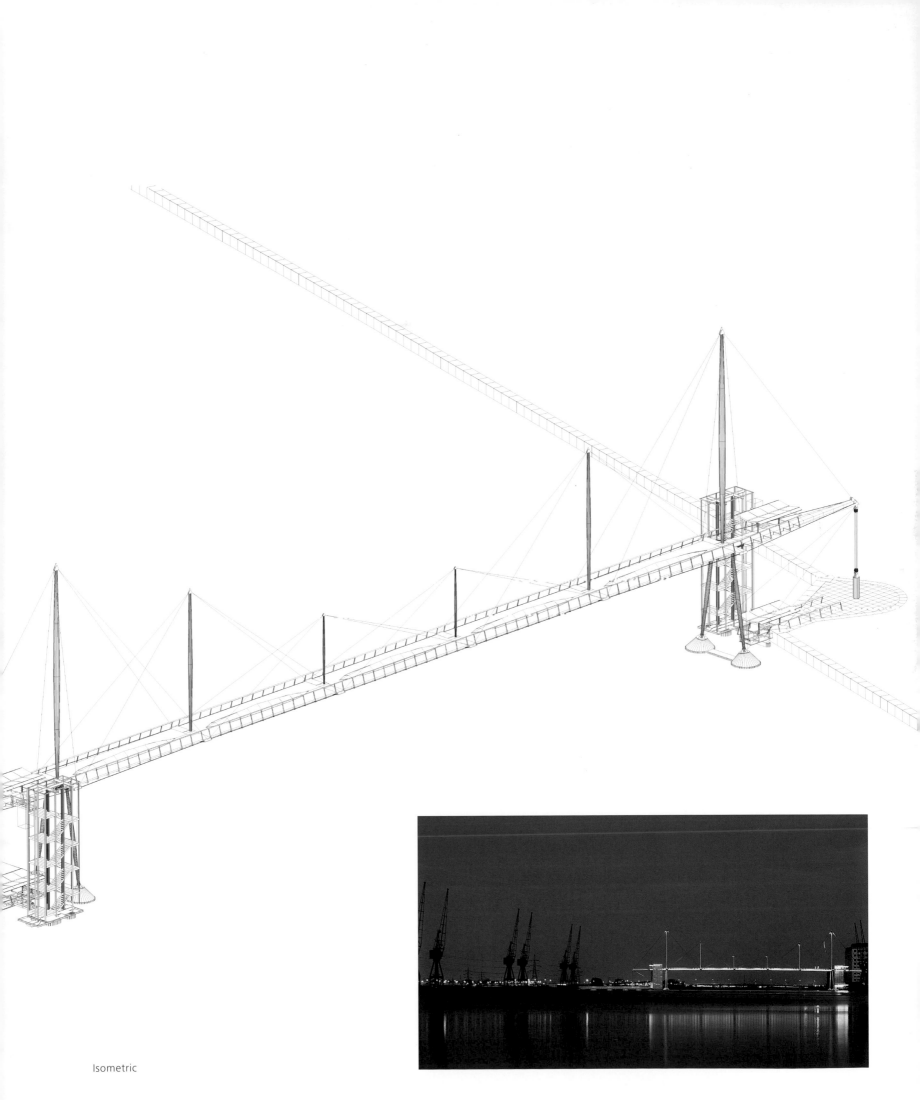

Isometric

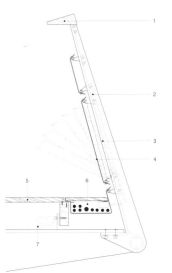

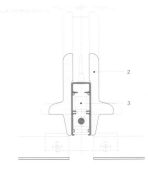

Typical handrail detail section:
 1 Iroko Handrail
 2 Cast aluminium balustrade bracket
 3 Balustrade light fitting
 4 Perforated stainless steel infil panels
 5 Iroko decking
 6 Perimeter service tray
 7 Soffit light fitting
 8 Monocoque panel beam structure
 9 People mover, power and guide track
10 Temporary maintenance gantry

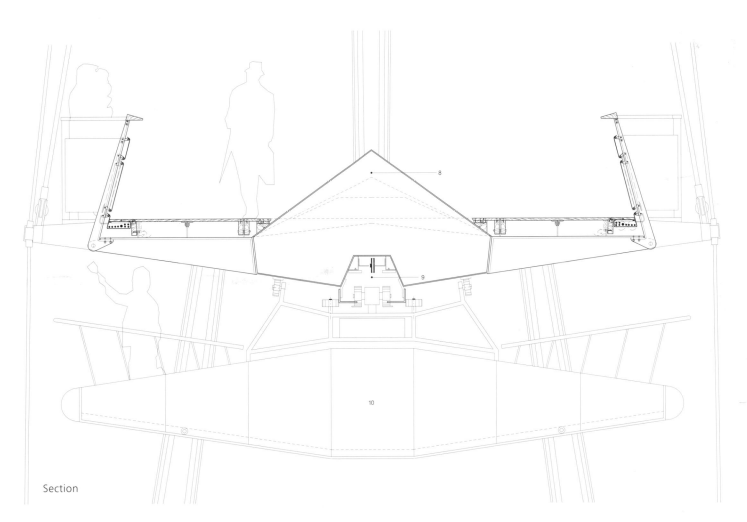

Section

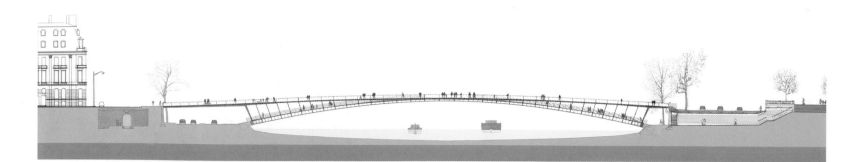

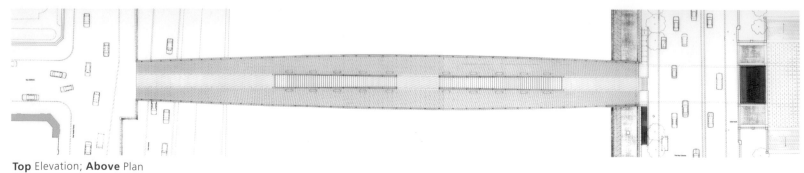

Top Elevation; **Above** Plan

SOLFERINO FOOTBRIDGE
Paris, France

As part of strategic upgrading of the Parisian infrastructure, the Solferino Footbridge over the river Seine links the Quai de France and the Quai des Tuileries. Replacing an existing structure, the new bridge was required to link the lower and upper levels of the quays on either side. Crossing the busy vehicular carriageway on the left bank, while connecting pedestrians under the roadway on the right bank, the bridge is a sophisticated solution to the establishing of important pedestrian connections in an urban environment dominated by vehicular traffic. The bridge is fully integrated within its context, and re-establishes the historical urban continuity between the two banks.

The bridge comprises two arcs. An outer form, of shallow radius, connects the high level quays, while a lower, tighter radius arc creates a stepped ramp connecting the lower levels. The two arches delicately coincide at the centre of the span, enabling pedestrian circulation between the different levels. The bridge and the crossing of the Seine are thus experienced in two ways – from above, open to the sky and addressing the expanse of the City, and from below, enclosed and intimate, relating to the water.

Two fabricated steel girders connected by cross-ribs form the primary body of the bridge. These not only offer a rigid structure but also provide a cage of steelwork at the lower level, enhancing the feeling of enclosure. The decking of the upper level and the steps is made from finely grooved hardwood, with handrails and benches on the deck formed of similar material.

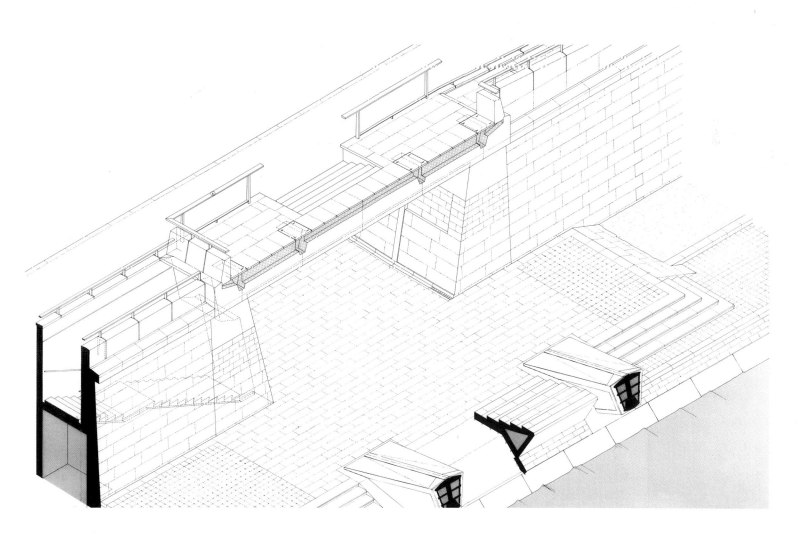

Axonometric

Section

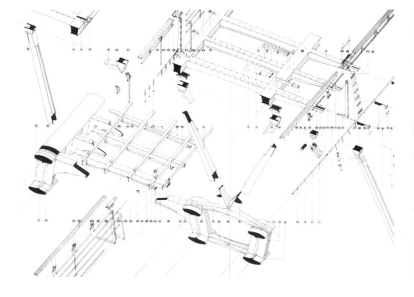

Top Section of handrail
Centre Section of deck
Above Axonometric of structure

BLACK DOG HILL BRIDGE
Near Calne, England

The bridge forms part of the North Wiltshire Rivers Route of the National Cycle Network. Crossing the A4 near Calne, it is set in a splendid landscape. The choice of a visually stunning structure celebrates and advertises the route to potential users, encouraging them to consider adopting bicycle transport.

The site has strong sweeping visual axes, and the bridge is designed to empathise with the nature and topography of the site. The sinusoidal curves are reproduced within the form, and the powerful asymmetry of the site also features strongly in the design. The final form of the bridge inclines asymmetrically on one side. The cross-section of the bridge, including the steel handrails, is also asymmetrical. The steel deck rises and cants to produce a self-steering velocity for cyclists. One key objective of the bridge was to disorientate the users, sensibly changing as many geometric variables as possible without becoming whimsical.

The bridge design results in a structure comprising two statically indeterminate compression arches and one tension tie arch. Each one counteracts the vertical, rotational, and out-of-balance forces present on this lightweight bridge structure. It is asymmetric and has a curved, inclined primary compression arch, with a rising footway curved in the opposite direction. This secondary arch provides stability and helps prop the main structure. The structural footway element is hung by inclined cables from the primary arch, and is supported on timber rocker beams. These beams, along with a rising rear tension tie arrangement, resolve the out-of-balance rotational forces resulting from the offset footpath – forces which are in part distributed by the handrail.

The primary arch is manufactured from treated redwood and curved, glulam beams, the maximum section size being 675 x 675mm by some 40m long. This choice of material was selected to harmonise with the green wooded landscape, subtly showing the impact of man on natural materials, and to provide efficient, cost effective, low maintenance, galvanised steel handrails and a galvanised steel deck.

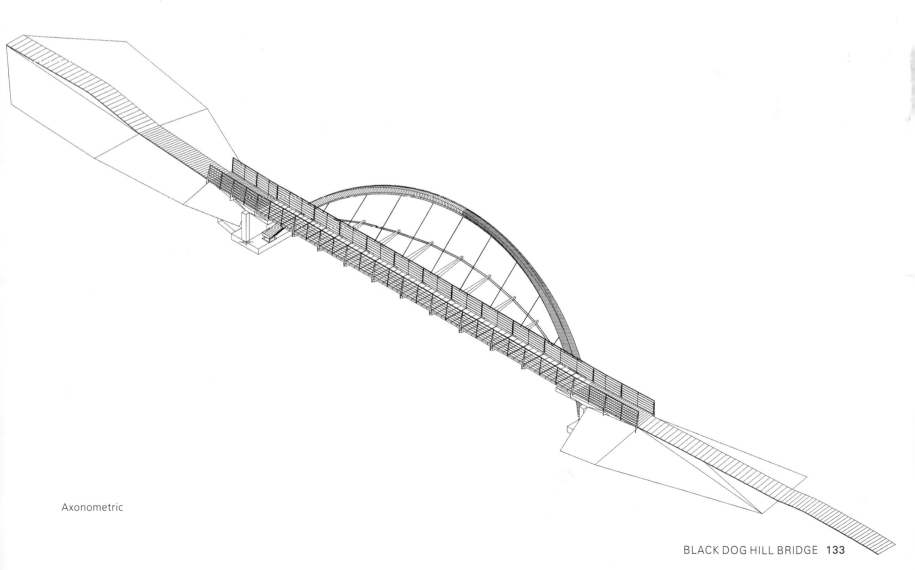

Axonometric

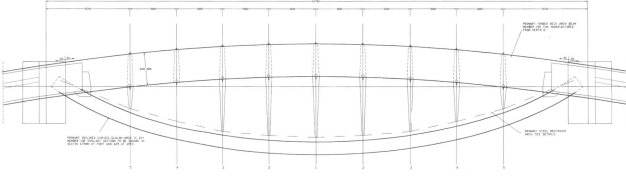

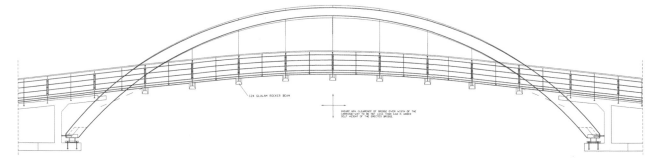

Left Plan and elevation
Right Exploded axonometric

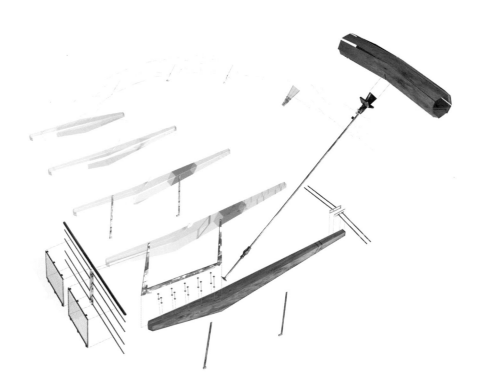

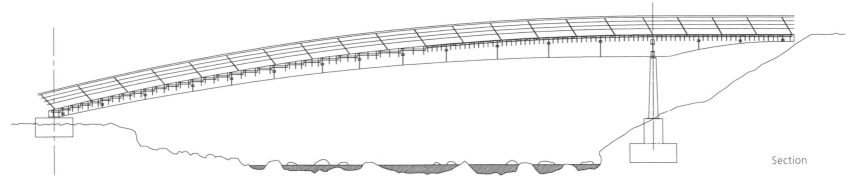

Section

RIVER IRTHING FOOTBRIDGE

Hadrian's Wall, England

The River Irthing Footbridge forms an integral part of the National Trail pathway along Hadrian's Wall. The Roman engineers' bold, simple, and direct approach to structure is clearly demonstrated by Hadrian's Wall. In this spirit the design focused on structure, rather than method and details, as the first step in ensuring the bridge would be sympathetic to its unique context.

Archaeology, ecology, method of construction, site access, and maintenance were all particularly critical in determining the design of the River Irthing Bridge. Coupled with the need for minimal intervention, the other major design factor was the difference in the levels of the two adjacent banks. To accommodate that change of level, in a way that would honestly respond to the existing topography, required the simplest and most direct structure. Taking the form of a single-span simple beam, the bridge rises gracefully from a flat meadow floodplain to the top of a wooded escarpment.

Crossing the River Irthing in a single 32m span, the bridge has a second 8m back-span which cantilevers over the steeply sloping west bank. The main beams are curved in shape, and taper from 1,100mm to 400mm at either end, eliminating the need for high abutments. The bridge is supported on two shallow concrete foundations.

It was important to minimise disturbance to this archaeologically sensitive site and the surrounding environment. Central to this concept was the idea of lifting in the main bridge beams by helicopter, thus avoiding the need to build access roads. These beams are now weathering naturally to a rich deep brown colour.

Plan

NICHOLAS LACEY & PARTNERS
ROTHERHITHE TUNNEL BRIDGE
London, England

Designed for both pedestrians and cyclists, the bridge fills an important gap in the cycle network north of the Thames. From a pair of slender outward leaning arches are slung suspension rods that support, by means of outriggers, a trussed deck. This braces the structure against the arches, resisting uplift as well as the live loads on the deck.

Along each side a stretched, woven filigree of stainless steel – a kind of transparent chain mail – protects the road traffic below from the threat of projectiles.

Flying free of the tunnel cutting, the bridge relies on just two points of support, thus minimising disruptions to the gardens on either side. Nicknamed 'the butterfly bridge', its bright steel form floats in the arcadian setting of St James's Gardens with its fine mature plane trees.

To minimise construction time, the bridge was designed to be assembled on site alongside the cutting. It was lifted into place in the space of one overnight closure of the tunnel.

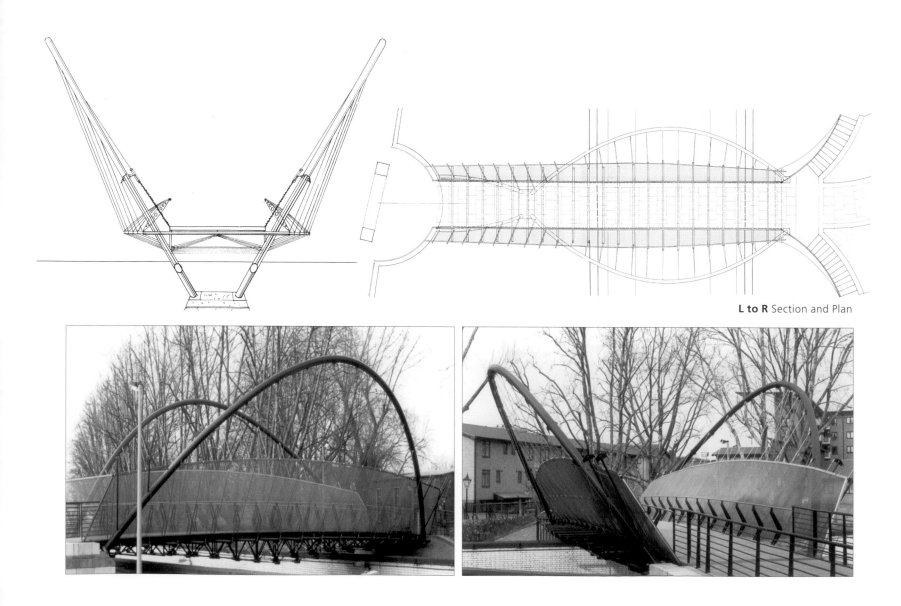

L to R Section and Plan

ST SAVIOUR'S DOCK BRIDGE

London, England

St Saviour's Dock footbridge fills an important gap in the riverside walk on the South Bank of the River Thames. It links Butlers Wharf and the Design Museum (just downstream of Tower Bridge) to the Mill Street area to the east, avoiding a long detour round the dock. Formerly the mouth of the Neckinger, one of London's 'lost' rivers, St Saviour's Dock is characterised by a number of Victorian warehouses, of four, five, or six storeys, many of which are listed buildings.

The design is a lightweight structure, a filigree of stainless steel, relating visually to the existing cranes and contrasting with the enormous masonry masses of the warehouses lining the dock.

The brief required that the dock remain accessible to small vessels, and the central span swings horizontally allowing passage to craft entering the dock. In order to keep structural member sizes to a minimum, the opening span is designed as a tension structure with just three pairs of compression members – the masts, the horizontal deck, and the handrail supports. All other structural members are tensioned rods. The structural principle is like an inside-out bicycle wheel, with the suspension rods being the spokes and with horizontal and vertical tubes in place of the wheel rim. The visual effect is one of transparency, a kind of stainless-steel spider's web.

The bridge is designed to be light enough to be swung by hand, weighing little more than a large saloon car. Yacht steering wheels positioned on each side jetty operate the hydraulic slewing motor, housed in the cone which surmounts the central support.

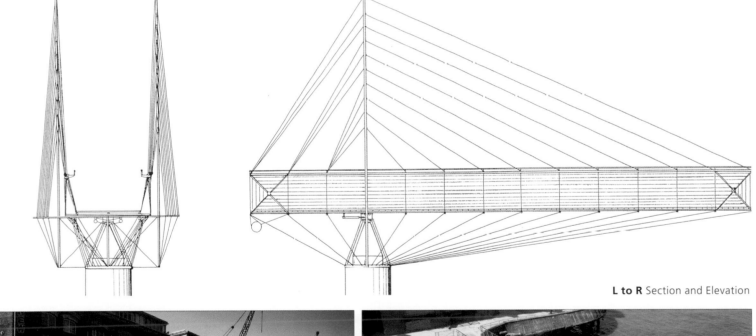

L to R Section and Elevation

MUSEUM PARK FOOTBRIDGE
Rotterdam, The Netherlands

Situated in the museum district of Rotterdam, this new bridge forms the central focus of a parkland setting. Connecting the Netherland Institute of Architecture (by Jo Coenen) and the Kunsthaal (by OMA), the park is conceived as five equal spaces with the new buildings located either end. Each is flanked by a hard, landscaped forecourt. These are connected by the new bridge across a central natural area of mature trees and planting.

The bridge forms a gentle, almost imperceptible slope, raising the pedestrian above the park and affording views into the canopy of trees and down into the field of ornamental shrubs and flowers below. This enables a new raised perspective of the flat topography for pedestrians, while at the same time preserving the landscape from being trampled underfoot.

The bridge deck is formed of black concrete, in stark contrast to the surrounding natural environment. The deck is perforated by small glass cylinders – in sunlight these cast patterns onto the landscape which resemble fallen leaves. In section the deck is very thin, and takes the form of a delicately undulating line drawn through the park.

The deck is supported below by a series of seemingly arbitrary arranged columns, inclined at contrasting angles and resembling the irregular nature of the surrounding tree trunks.

This most delicate of bridges establishes an important route, gently raising pedestrians to provide an elevated view of the mature landscape, while at the same time conserving the landscape's natural beauty.

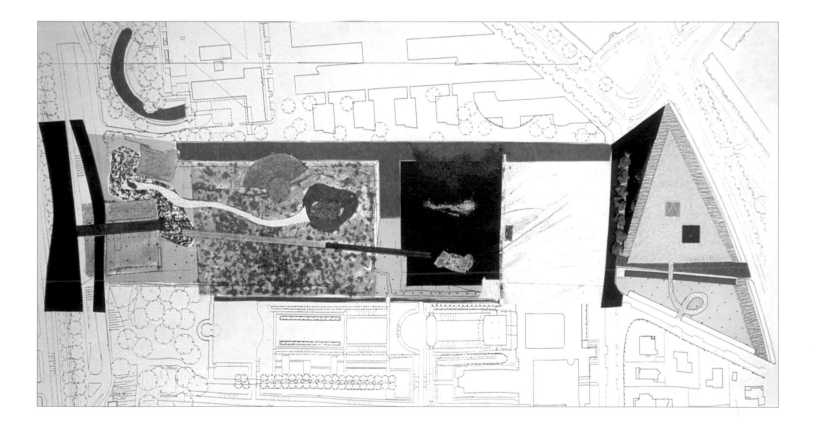

Site plan

Perspective

MIHO MUSEUM BRIDGE

Shigaraki, Shiga Prefecture, Japan

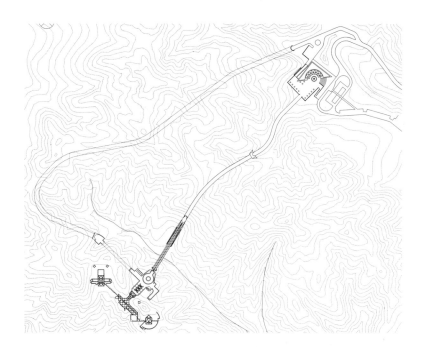

The Miho Museum is situated between two ridges of a nature preserve, built into a precipitous mountainside and accessible only via a tunnel and bridge.

Visitors arrive by car or bus at a triangular reception pavilion which is articulated on the exterior with Japanese stucco and tile, and which houses ticketing services, a gift shop, and a cafeteria within. From there, guests travel through the mountain tunnel, either by foot or transported by a small electric car. The mouth of the tunnel opens onto a 120m post-tension bridge. This spans a precipitous drop, and leads visitors directly onto the museum plaza.

Contoured site drawing　神慈秀明会美術館計画

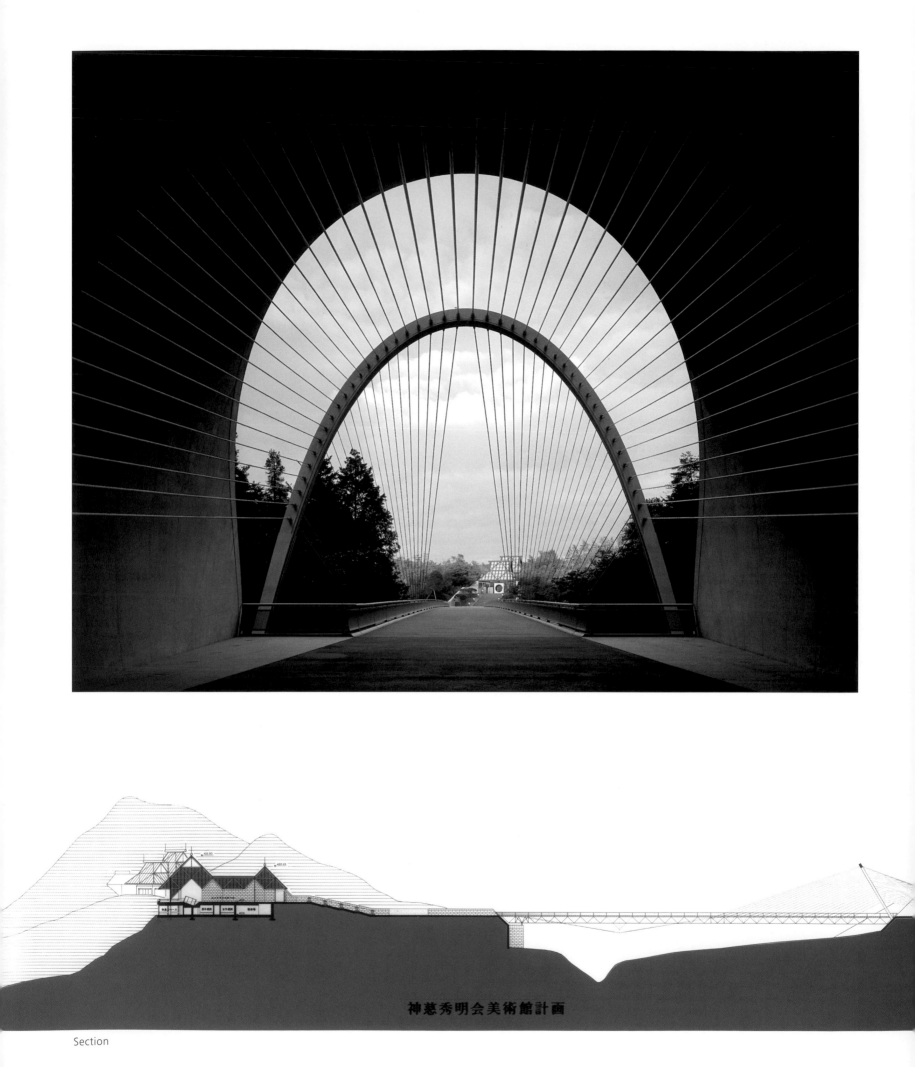

神慈秀明会美術館計画

Section

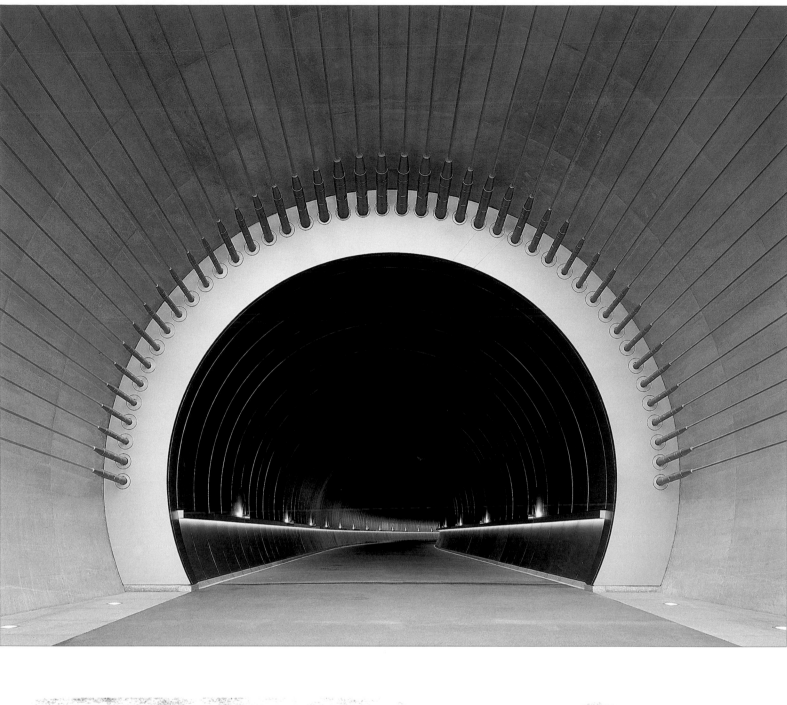

Isometric

BERCY-TOLBIAC FOOTBRIDGE

Paris, France

This new pedestrian bridge links the Tolbiac and Bercy neighbourhoods, crossing the River Seine between the François Mitterand National Library and the Bercy Park. The bridge reaches across the river in a single, continuous span, unbroken by intermediate supports.

With a free central span of 190m, the full length of the bridge is an inseparable blend of architecture and engineering. It is an efficient structure, which combines spatial potential with lightness and strength through the synergy of two collaborating elements, a remarkably slender arch balanced by a pretensional, suspended catenary. Three parallel decks follow these two curves, the central deck rising with the arch while the flanking decks fall with the catenary. The gentle slopes, less than four per cent in general, allow for easy access.

Arch and catenary are tied together by a series of 'obelisks' – vertical tapered struts springing from the arch – to form a semi-vierendeel beam. A support system, fixed at each extremity and articulated at the quarter points of the span, is employed to reduce the required inertia of the composite beam. The fixed end supports are assembled from a pair of vertical tension rods restrained by a pair of bent struts, which distribute loads from the arch to its foundations.

The pedestrian bridge draws its elegance and finesse from the harmonious unity of its structural and spatial functions. Its use of contemporary materials and construction methods continues the tradition of technical modernity expressed by many historic Parisian bridges.

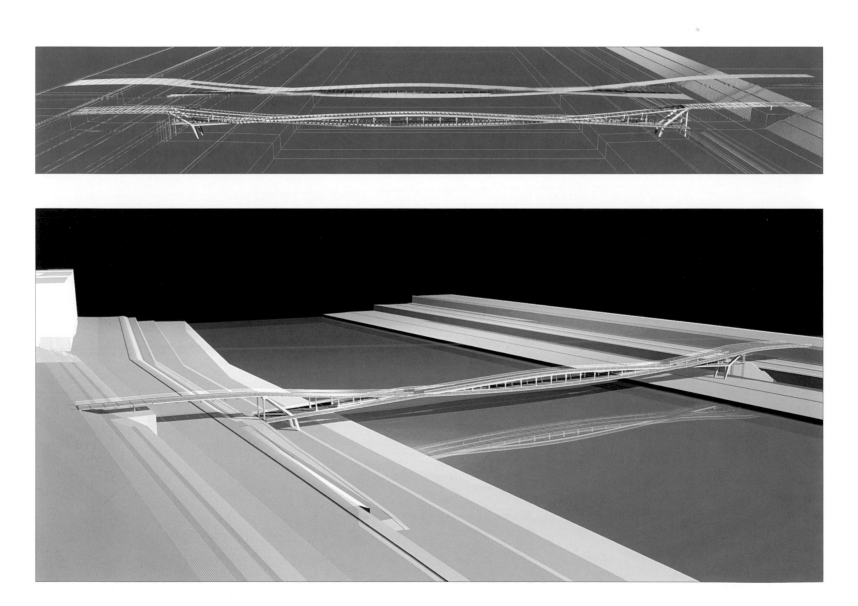

SKYWALK
Hanover, Germany

The skywalk is designed to give weather protection to the moving pavements linking the railway station to the main entrance of Hanover's Expo 2000. The slender 340m long gallery stands alone, elevated some 6m above the ground, resembling a space-age transporter temporarily immobilised on its stilt-like legs.

To maximise views through the curved façades, the primary structure is restricted to zones located above and below the moving pavement, where binocular section fuses into a single element. The steel framework is opaque and deliberately massive, highlighting by contrast the lightness and transparency of the facades themselves. This contrast was reinforced by the decision to transfer roof loads directly onto the supporting columns, located at 30m centres, rather than via the ribs of the facade – a choice which rejects structural optimisation in order to reinforce an architectural idea.

The pronounced curvature of the flanks enables shell elements to be created from glass panels. This allows an increase in the span to 2m without the need for intermediate transoms, and to reduce the size of the steel ribs themselves. The ribs are treated as arch elements, fixed at the floor and roof levels and laterally stabilised by the glazing; they work essentially in compression rather than in bending.

Spared from all the primary load transfer, the ribs achieve an extreme lightness, fluidity, and immateriality of the facade, a sensation created by displacement of mass rather than an absolute truth.

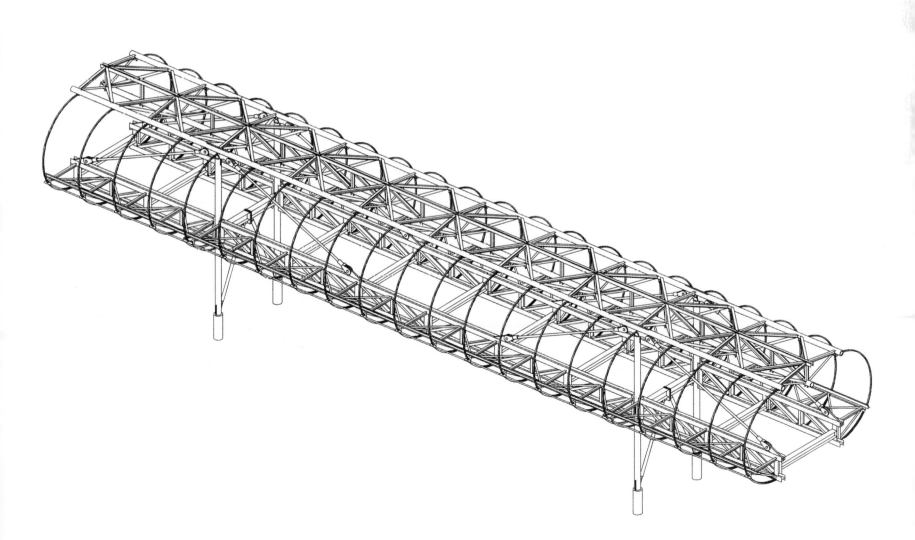

Axonometric

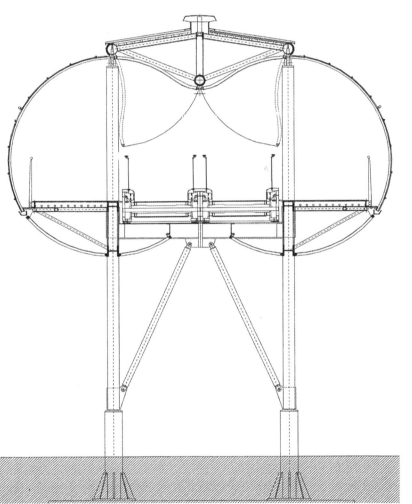

Section

JAPAN BRIDGE

Paris, France

The Japan Bridge crosses a busy road to form a link between two parts of the newly developed area of Valmy at La Défense, the business and financial district to the west of Paris. The footbridge spans 100m, some 15m above road level, over a total of seven lanes of traffic.

The initial idea for the bridge came from Kisho Kurokawa, architect of the Tour Pacifique, one of the two office blocks on which the bridge is supported. He drew on the traditional Japanese arched bridge design, where a walkway follows the curve of the arch. In consideration of the inability of the supporting buildings to take the horizontal thrust of an arch, RFR developed this idea into the structural solution of a double-tied arch. The two parabolic arches, made of welded steel triangular hollow section, lean inwards and converge at their apex, stabilising each other and providing a wide base which gives the bridge resistance against overturning. Each is tied by its own 'bowstring' tendon, a solid 200mm diameter steel rod, to which it is connected via a series of hangers. Pre-

stressed by the weight of the structure, these create a stiff, stable plane between arch and tendon.

At the intersection points of tendon and hangers are the main nodes, which also form the connections with the deck support structure. The gently curving deck, of trapezoidal pre-cast concrete panels on a horizontal steel truss, is covered by a curved glass enclosure. Supported above the tendons by a system of braced struts, the glazed walkway remains separate from them, allowing them to appear as slender tension elements.

The apparently simple structure required complex analysis to investigate the buckling behaviour of the arches and the total torsional stability. In addition, due to the asymmetrical geometry of the supports, every element has a different length and every angle in the structure is unique. The use of computer modelling was required for geometrical and structural analysis, along with fabrication information.

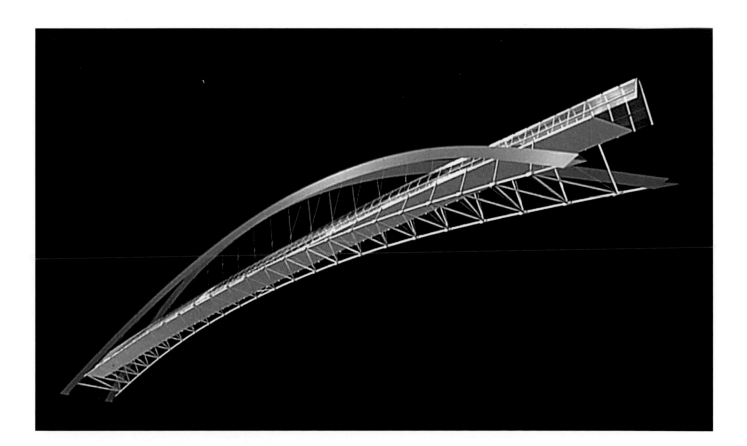

Computer model

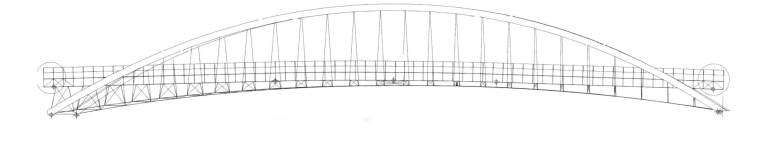

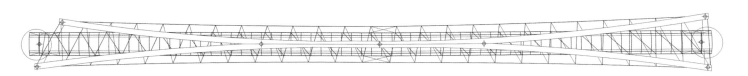

Elevation and plan

Node detail

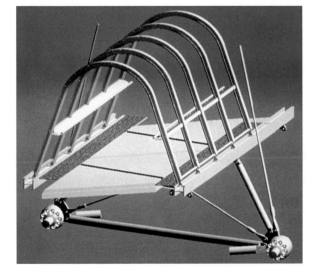

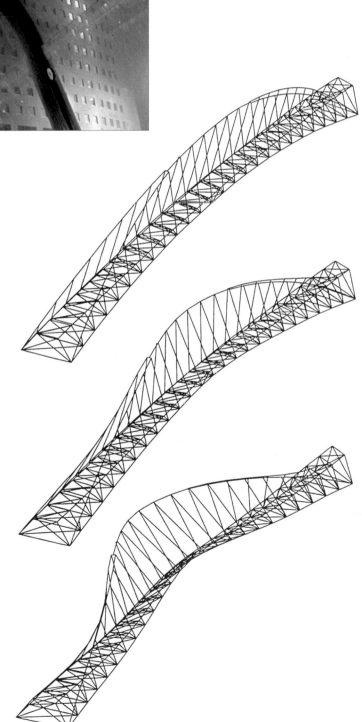

Above Constructional detail
Right Computer test diagrams

SCHLAICH BERGERMANN & PARTNER
CONVERTIBLE BRIDGE OVER THE INNER HARBOUR
Duisburg, Germany

Located north of Cologne is the Rhine harbour town of Duisburg. This bridge links the city centre with a new development across the harbour. Due to ship traffic, the bridge design had to allow for a height of 10.6m above high water level. The relatively short span of the harbour, combined with this height requirement, would have posed considerable obstacles to the desired ease of pedestrian crossing. However, the radical design resolves this by creating a footpath that can be raised and lowered.

Designed as a back-anchored suspension bridge, the structure is 3.5m wide with a span of about 73m. To enable movement, the back-stays can be shortened by 3m with the use of hydraulic cylinder jacks. Thus the masts pivot outwards, and the deck, consisting of pre-cast concrete elements, is automatically raised into an arched shape. To enable the bridge deck to curve during lifting, the deck elements are joined by hinges. The additional length of the walkway is provided by one extra deck element on either side, which can be pulled out from pockets in the abutment.

Watching this in process presents a wonderful, theatrical experience.

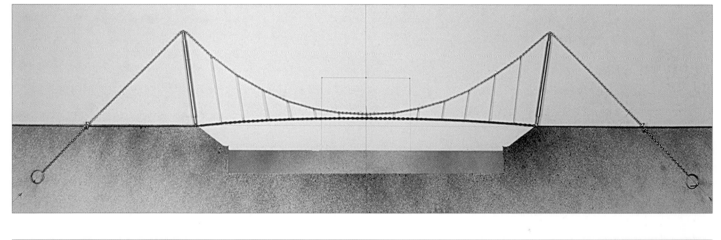

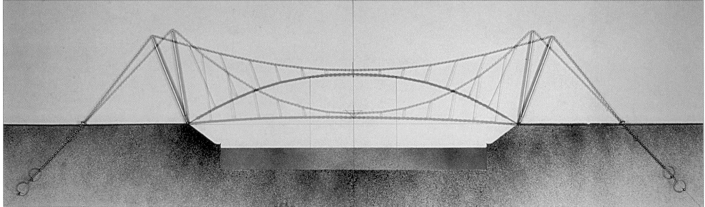

Sections

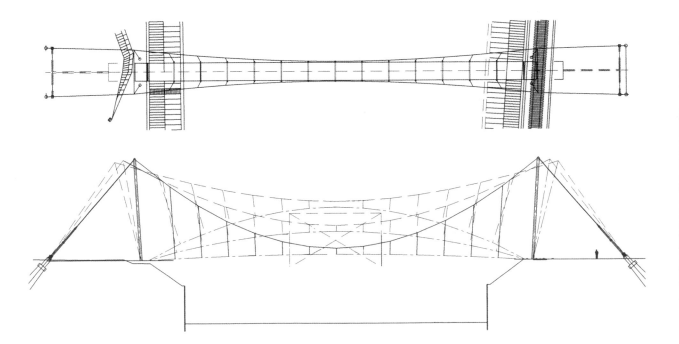

Plan and section

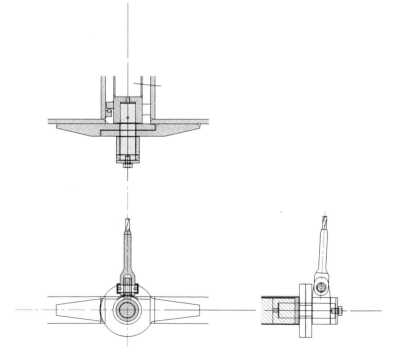

Connection detail

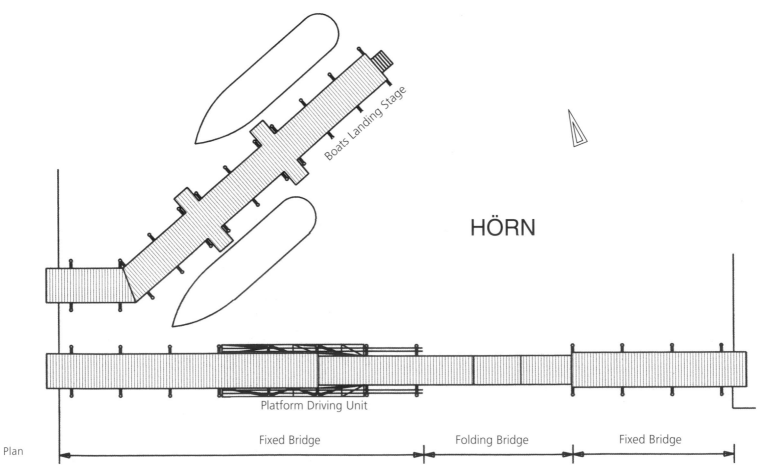

Boats Landing Stage

HÖRN

Platform Driving Unit

Fixed Bridge

Folding Bridge

Fixed Bridge

Plan

FOLDING BRIDGE OVER THE FÖRDE

Kiel, Germany

The bridge is located near the harbour of Kiel and reflects the surroundings of industry, ships and cranes. An important element of the large-scale regeneration project for the area, this pedestrian link was required to connect two districts on either side of the harbour.

The bridge takes the form of three sections hinged together, which cross the 22m wide span. Requiring the passage of large vessels through the harbour, these hinged sections can be folded back to provide a clear through-way for shipping.

In the closed position the bridge appears like a conventional single-direction, cable-stayed structure, with a 5m width. However, when required, a cable system allows the deck to be folded back rapidly, a process which occurs approximately 10 times every day. In all positions and under all loads, especially under wind loads, the cable system is statically determinate. No hydraulics or springs are used to keep the cables under tension. The whole bridge is moved by the continuous rotation of one single-speed winch, on which all cables are coiled, with the winches driven by hydraulic motors.

This extraordinarily dynamic structure presents a unique solution to the problem of allowing for frequent pedestrian crossing on the level while accommodating the needs of the working harbour.

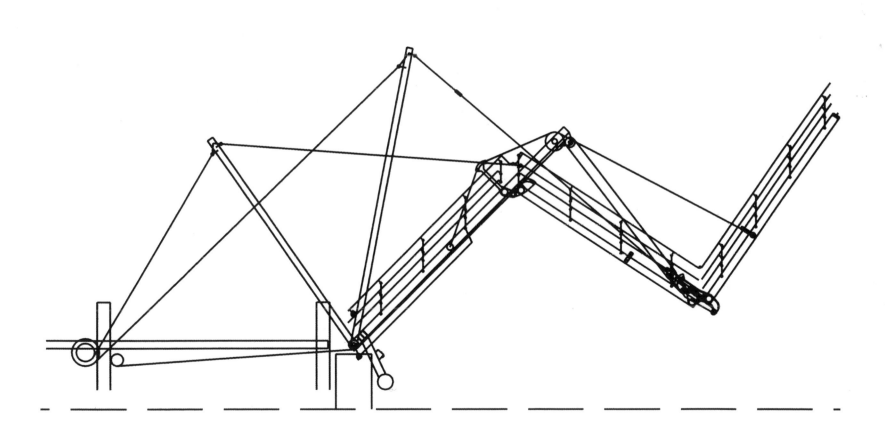

Elevation

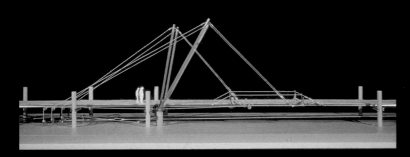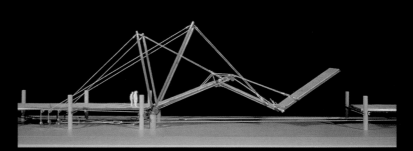

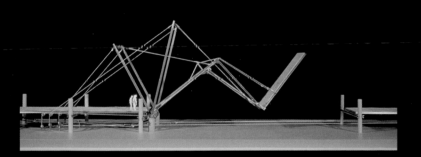
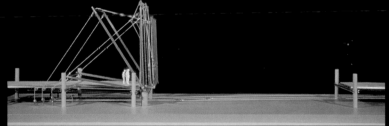

Model showing opening and closing mechanism

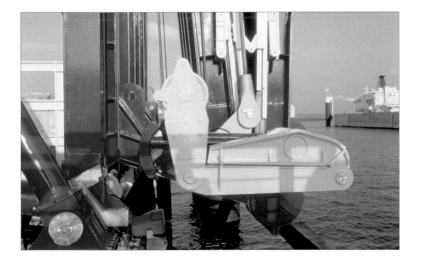

PEDESTRIAN BRIDGE OVER THE NECKAR RIVER

Near Max-Eyth-See, Germany

This 114m-span pedestrian bridge is close to Stuttgart, in a beautiful and unspoilt natural landscape. The light and translucent suspension bridge was designed to fulfill the specific environmental requirements of such an area.

After comparing different solutions, a back-anchored suspension bridge was chosen. Its girder can be freely shaped in plan, and thus be adapted to the specific topography – a steep hill with vineyards on one side and a flat park with beautiful trees on the other. Prefabricated concrete slab elements were lifted from barges to be suspended stepwise from the main cables installed earlier.

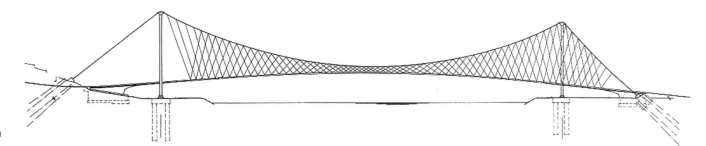

Section

Site plan

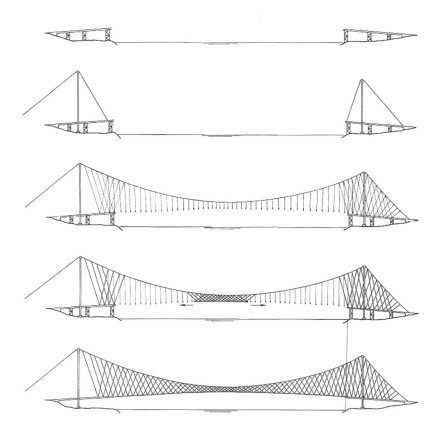

Construction stages

PEDESTRIAN BRIDGE OVER THE WESER RIVER

Minden, Germany

This bridge takes the form of what is, in essence, a classical back-anchored suspension bridge, since the requirement was that it should be as light and transparent as possible. This was to avoid obstruction of the view down the river Weser, with the Porta Westfalica monument in the background.

With a beautiful park on one side of the river, and an open space on the other, the two banks of this site are not parallel. A straight bridge would, therefore, have been misaligned on one of the banks. With the deck curved in plan, however, it adapts homogeneously to this specific natural boundary condition. The inclined masts reflect this curvature and also lend this small bridge its own unique quality of form.

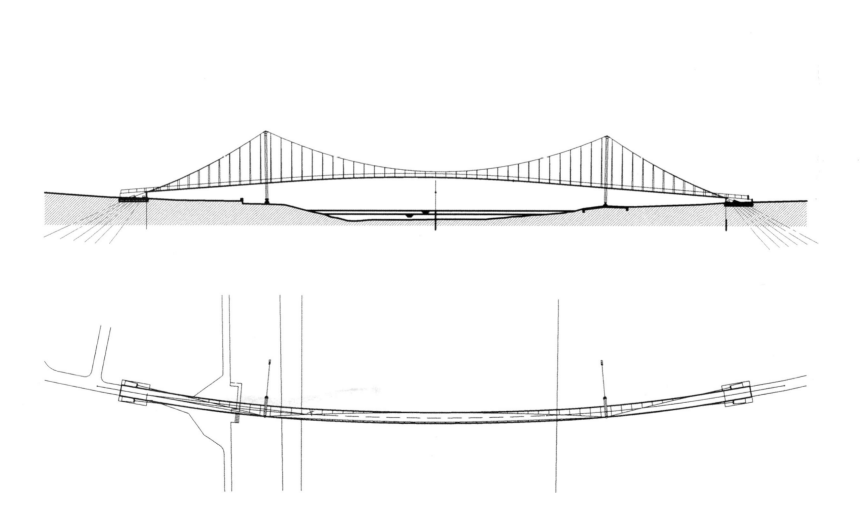

Elevation and plan

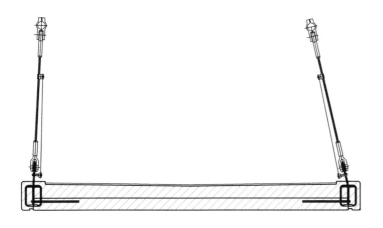

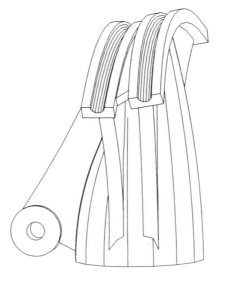

Above Deck section
Right Detail of saddle

ORACLE BRIDGES
Reading, England

Whitby Bird & Partners won the competition to design two new public footbridges and a canopy, one of the largest installations at The Oracle, a new retail and leisure centre in Reading. The development was intended to express the town's successful identity as a thriving centre for new technology.

The bridge at the western end of the development spans the River Kennet, providing a direct link between the Riverside car park and the Rotunda entrance to the new centre. The second bridge is an elliptical, cambered steel tube structure, whose span forms a single 58m long x 4.6m wide sweeping curve. Both bridges feature aerofoil profiles. The bridges and the canopy structure are arranged to create a landscape arena and remain true to the original concept, which brought together the disciplines of architecture and engineering.

As a compositional element the curved bridge is juxtaposed with the straight bridge – however, the elliptical nature of the space dominates, reinforced by the form of the canopy. The bridges are integrated with the landscape, encouraging people to linger in this new civic space. The enveloping arms of the elliptical bridge not only appear to embrace the site, but also provide a curved timber seat as a resting place for people to sit and enjoy the spectacle.

The structural ingenuity is modestly apparent, allowing the sculptural forms of the bridges to be clearly understood. The structural concept of the bridge is inspired by the efficient form of a modern aircraft wing, which is slender to minimise drag, yet immensely strong with its structural plate elements top and bottom. These form a closed box, subjected to twisting forces due to the changing load conditions.

The use of oak decking makes reference to the site's origins as a cooperage. The timber also provides the visual richness of a warm, tactile material in contrast to the modernity of the continuous translucent glass strip at the edge of the deck. At night this is illuminated with a blue light, and heightens the extreme slenderness of the bridge form.

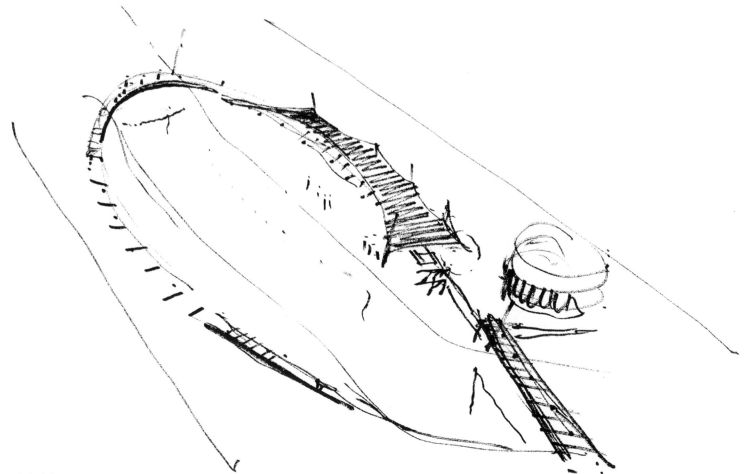

Concept sketch

Section

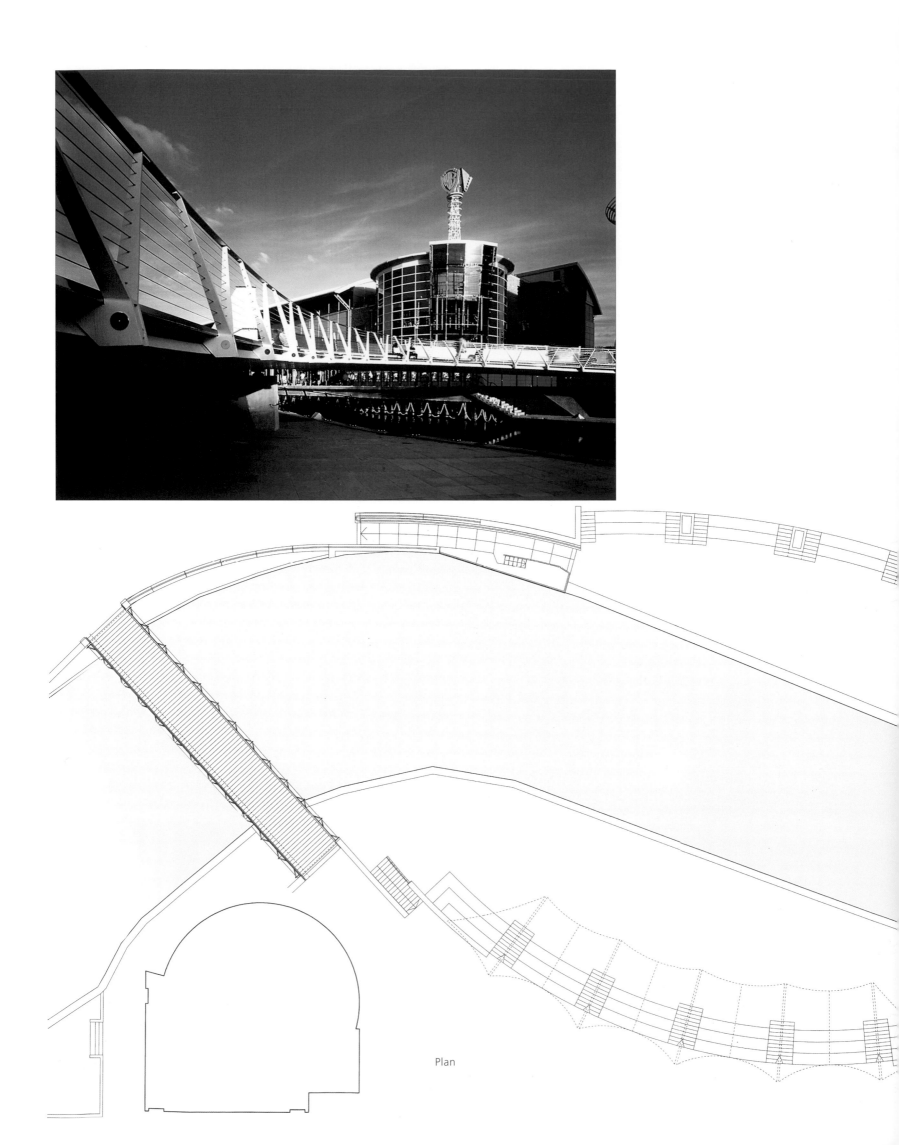

Plan

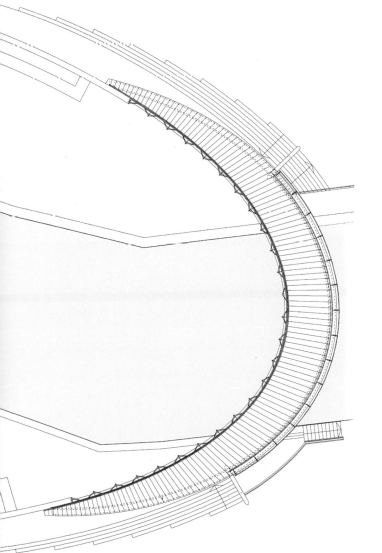

RIVER LUNE MILLENNIUM BRIDGE

Lancaster, England

The new bridge over the River Lune at Lancaster reinforces the sense of place and visual containment of the historic quay, which had been eroded by the loss of the previous medieval bridge.

The shape of the bridge is that of a 'Y', linking the north bank to both the quay and the old railway viaduct, and allowing easy movement of both pedestrian and cycle users crossing the river between Lancaster and Morecambe.

The twin-masted, cable-stayed structure takes advantage of better ground conditions on the south of the river, while evoking images of the tall ships that occupied the quay in the seventeenth and eighteenth centuries. The gangway leading down to the quay reinforces the maritime image, and acts as a brace against the sway of the masts.

The main deck of the bridge is a 600mm-deep steel box girder, fabricated from stiffened plate. The masts are tubular in form, and fabricated from bent plate tapering from 1,200mm diameter at their centre to just 150mm diameter at their feet, where they are standing in stainless-steel castellated shoes. These shoes are connected to a steel casing supported on an elliptical concrete pier. Cable stays support the span, and the triangulations formed by the tie-down cables on each side maintain the stability of the masts.

The gangway deck is constructed using perforated aluminium planking, allowing the passage of light on the supporting structure during the day and LED luminaires to filter up at night. The use of aluminium also reinforces the difference between the structure of the new deck and the gangway.

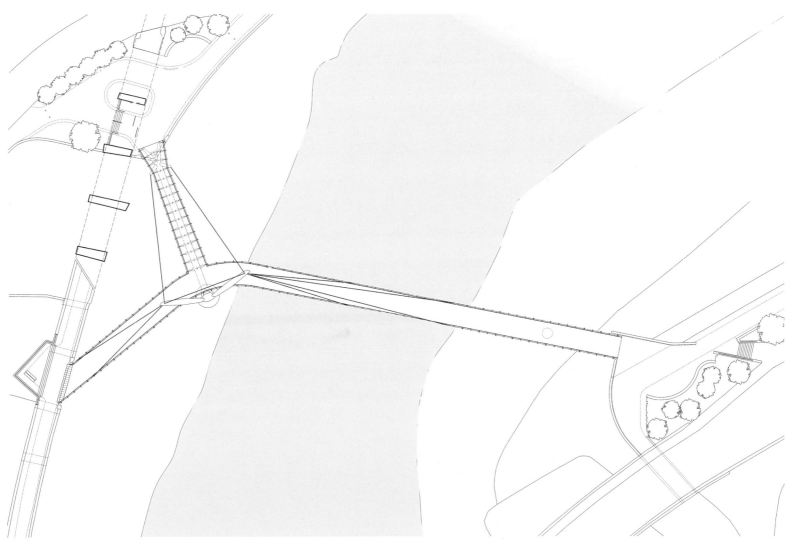

Site plan

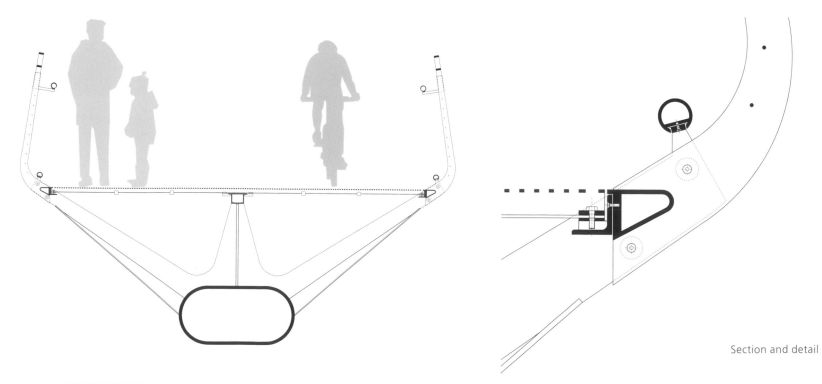

Section and detail

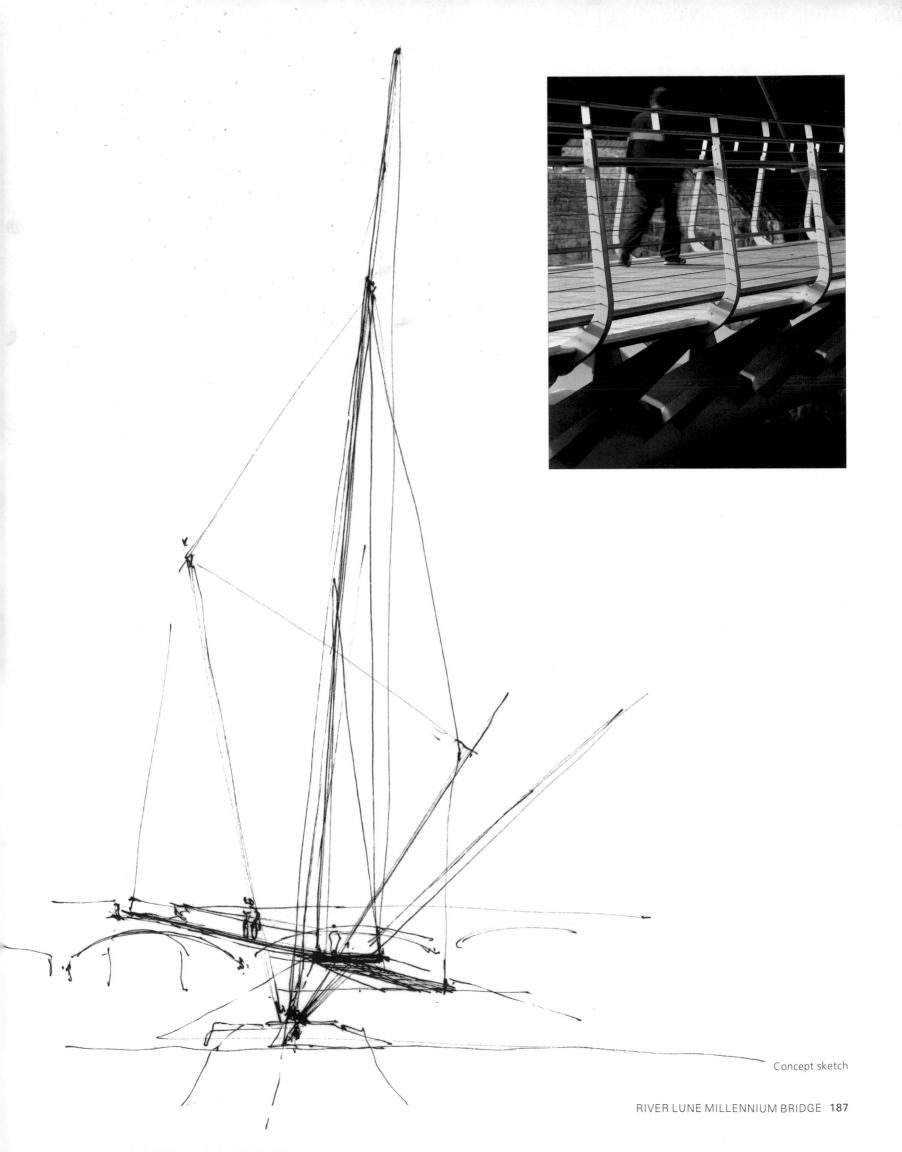

Concept sketch

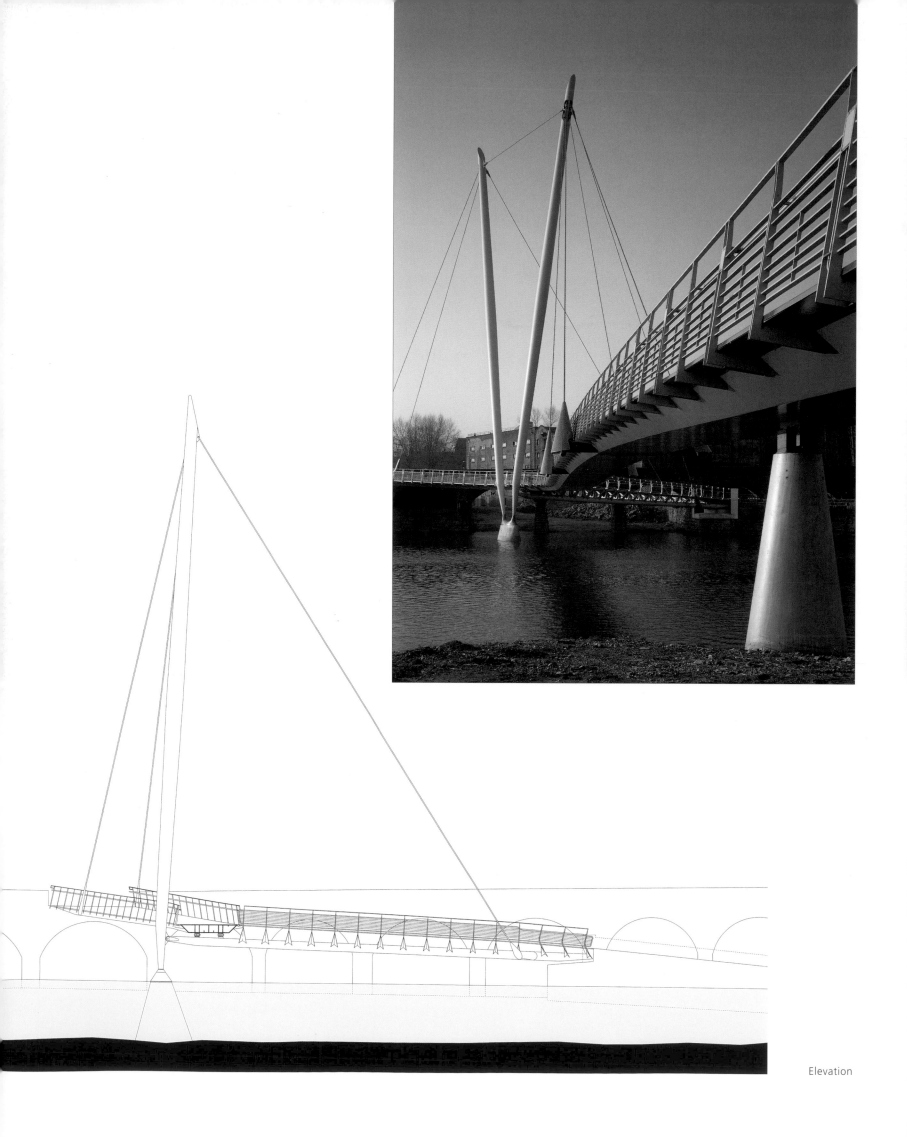

Elevation

SHANKS MILLENNIUM BRIDGE

Peterborough, England

This bridge is part of Green Wheel, a millennium project in Peterborough, providing a pedestrian and cycle route to encircle of the city with 'spokes' running into the centre. Shanks Millennium Bridge is the final link of the route, crossing the River Nene to the east of the city.

The design is a conscious response to the different natures and needs of the bridge's users, and emerges from the site in a very natural form. The concentric curves are a product of the existing levels on the site, forging a route across the river while simultaneously creating an intimate sense of place and opening out to the landscape.

Due to the need for a sturdy and enduring construction, appropriate for the carriage of horses and to the Fenland landscape, inspiration was taken from the Corten sculptures of Richard Serra. Corten is an elemental material which matures to become one with the landscape, and responds to the curve of the bridge by weathering in gradations of colour according to its orientation to the sun.

The main structural element of the bridge is conceived as a pair of 'folded arms', which offer the cantilevered pedestrian walkway on the inner rim of the curve. The Corten box forms the bridleway at a minimum width of 2.5m. The bridleway is contained on the outside by an inclined timber screen, which provides the mandatory 1.8m high parapet, whilst shielding users from the wind and avoiding silhouetting them against the skyline which would disturb the nesting of the environmentally sensitive river birds. The timber pedestrian walkway (with a continuous width of 1.6m) is raised up from the bridleway, visually reducing the mass of the structure and lifting the pedestrians relative to the horses and associated parapet. The Corten box also forms the structural spine of the bridge, resisting the bending and torsion generated by the loading. The bridleway has an in-situ concrete deck surface, providing the durability and solidity required for equestrian use whilst also helping to dampen potential live movement within the structure.

The timber deck is cantilevered from the front of the box section by slender Corten steel ribs, creating a distinct structural element. The ribs cradle the lightweight front deck section, and continue upwards to form elegant handrail uprights. The sculptural nature of the bridge is reinforced by the design of each of the Corten piers. They are composed of slender, ribbed, structural boxes, the ribs not only accentuating the elegant form but also expressing load distribution. The angle of the two main piers either side of the river maximises the width of navigation, and heightens the dynamic of the curve which appears to float over the river.

Plan

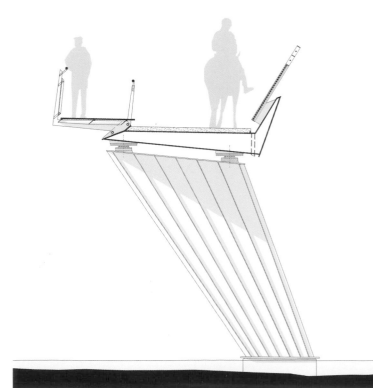

Section

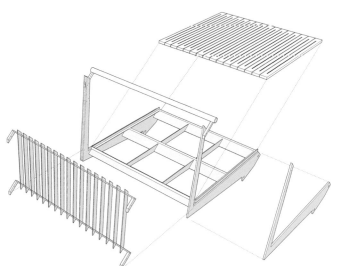

Detail of deck

The York Millennium Bridge exemplifies advancements in engineering design and quality in construction. This ethereal bridge is the focal point of York's commemorative millennium projects and is located at the transition between the city's formal riverside boulevard and the ancient flood meadows to the south of the city.

The winning bridge design brings together two separated communities with the creation of a piece of contemporary architecture and engineering, in turn providing a striking new asset to the historic landscape. Linking the south-east and south-west parts of the city, as well as the University on the east bank and housing on the west, the bridge is a key part of the National Cycle Network, built in response to the strong demand by the people of York to be able to move around safely, separated from vehicular traffic.

The bridge is strikingly asymmetrical. The stainless-steel arch describes a segment of a circle, which is set at an angle of 50° to the horizontal and so lies back towards the south and the wilderness beyond. This circle represents a bicycle wheel that would have 365 spokes if it were complete. The arch cables radiate from a central point within the earth, and set about stabilising the arch just as the spokes of a wheel stabilise the rim. In place of the hub of the wheel is the bridge deck, constructed predominately from carbon steels.

A timber seating area, running the entire length of the southern central span, is incorporated into the bridge design to encourage people to rest while experiencing unique views towards the city centre. The structure is also lit subtly, with special features illuminated to accentuate key elements of the fine architectural detailing.

Having overcome numerous technical challenges, the site was besieged by the worst flooding York had ever experienced. The project was eventually completed on 10 April 2001. This premium quality structure represents the efforts of a broad spectrum of professionals who contributed to the project.

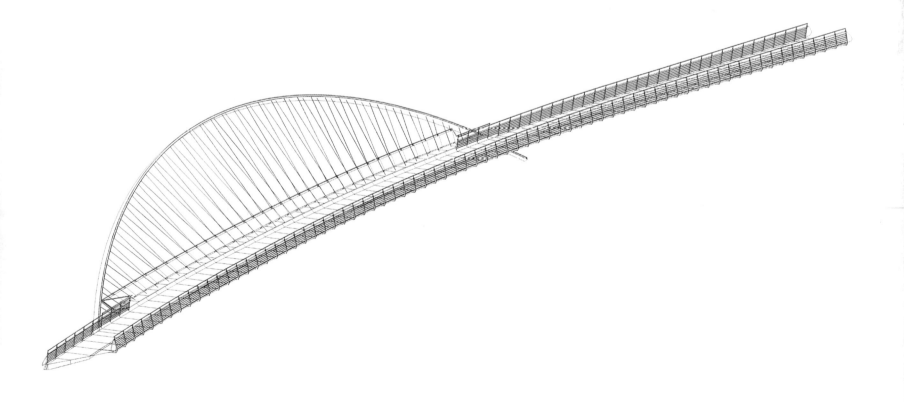

Axonometric

Plan

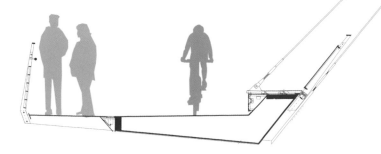

Section

BUTTERFLY BRIDGE

Bedford, England

This is the winning entry in an open competition set to procure an appropriate landmark and pedestrian crossing over the River Great Ouse in the town of Bedford. The bridge is set on the upper river, spanning 30m between willow-lined grassy banks, and adds to a series of crossings upstream from the historically important Bedford Suspension Bridge. An adjacent boathouse animates the river with the competitive rowing traffic which characterises this stretch of water, most evidently during the bi-annual Regatta during which over 500,000 people crowd the banks and bridges for a single weekend.

The new bridge translates the basic system of the innovative Bedford Suspension Bridge into an evolutionary design applying contemporary solutions and material capabilities. The twin-arched trusses of the historic bridge are revisited in the form of high arching parabolas, each of a single, circular, hollow steel section. The arches emanate from a single point on each bank and are canted to create an opening arch form, horizontally

16m from crown to crown, which alludes to a vision of butterfly wings. The composition is thus vaguely organic, like some mechanical insect on the flood meadows. This sense, in addition to the references to its neighbour, locates precisely the bridge in its setting.

Beneath the arches, a shallow arched timber deck is slung via rod hangars connected to structural stirrups, combining a deck support member with structural balusters. The timber deck then spans, self-supporting, between the stirrups. The suspended deck terminates as it passes through the split arches and adjoins a landing on each bank, cantilevered from an expressive concrete footing to the arch springings. Balustrades, leaning out from the deck, combine with a steel footplate covering longitudinal deck lighting, to encourage spectators to use the bridge as a platform to view rowers below. During the River Festival the bridge temporarily becomes a grandstand.

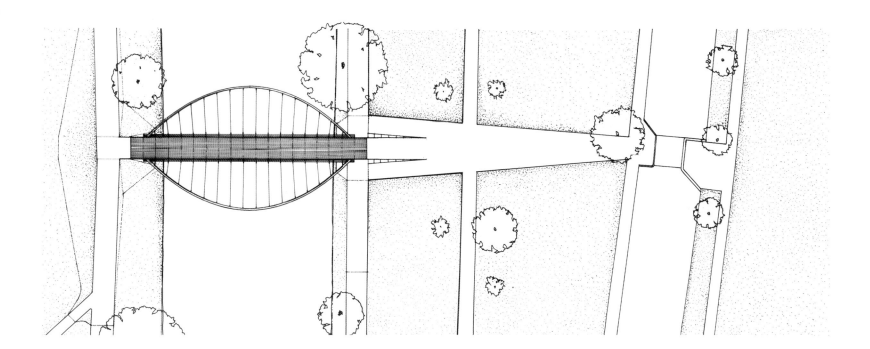

Site plan

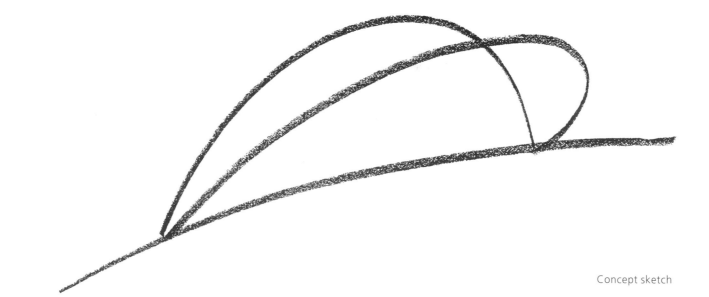

Concept sketch

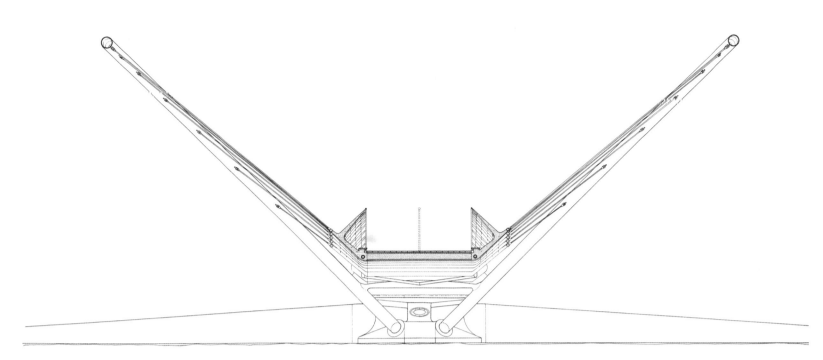

Section

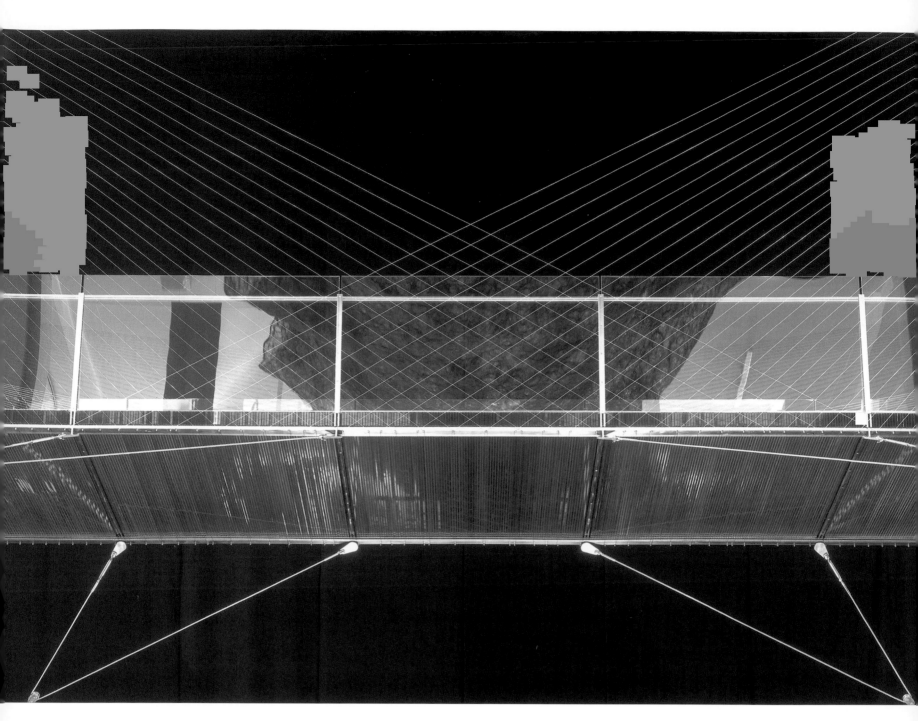

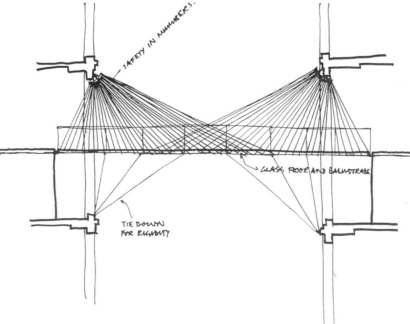

Concept sketch

CHALLENGE OF MATERIALS FOOTBRIDGE

London, England

At the heart of the Challenge of Materials Gallery, a major new gallery at London's Science Museum, this dramatic and innovative new footbridge spans the atrium at mezzanine height. It floats across the space to provide a visual focus and a marker for the gallery from the adjacent areas. Comprised of few elements and limited material, the bridge is minimal in the extreme.

Designed to the limits of technical feasibility, it offers a clear demonstration of material capability. Aiming to be almost imperceptible, the bridge comprises a deck formed by 828 abutted glass planks, suspended by an array of ultra-fine stainless-steel wires. By channelling these wires through stress gauges, and incorporating acoustic devices and lighting systems devised by the artist Ron Geeson, the bridge becomes an interactive exhibit, challenging materials to the limit and enticing visitors to cross.

The self-weight of the bridge is about 8.5 tonnes and the maximum people load 12 tonnes. This is all supported on just 40kg of wire, 1.58mm thick in total and about 2.6km long. The vertically laminated planks of ordinary glass are 19mm thick – the first use of this type of deck. The wires are collected at four points, and the forces pass through eight load cells that monitor the forces. These and other sensitive strain gauges are amplified to animate the bridge with responsive light and sound.

The Challenge of Materials Gallery is a key museum site – it exploits fully its position immediately above the great nineteenth-century beam engines and steam pumps near the main entrance by means of a scintillating glass bridge, held aloft by fine steel wires, which spans across the atrium void. The bridge is the breathtaking centrepiece. A visible demonstration of the innovative use of materials, it also performs the highly practical function of allowing visitors to circulate freely around the new gallery space. And, by its very juxtaposition with the massive machines below, it emphasises the immense advances that have been made in the science of materials and technology.

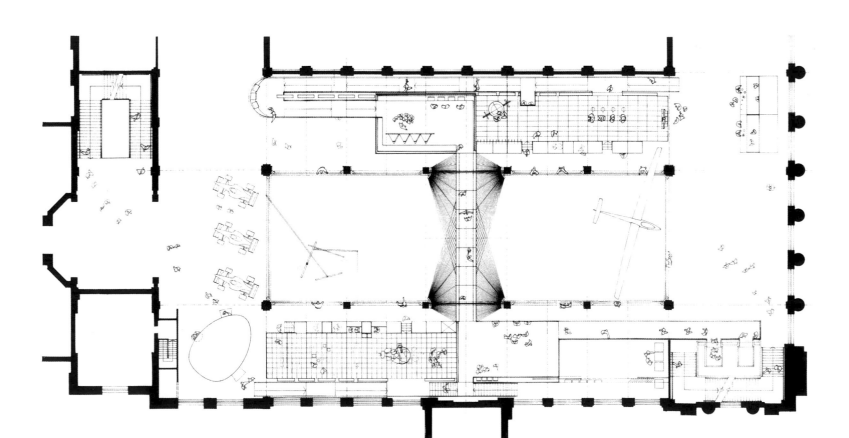

Context plan

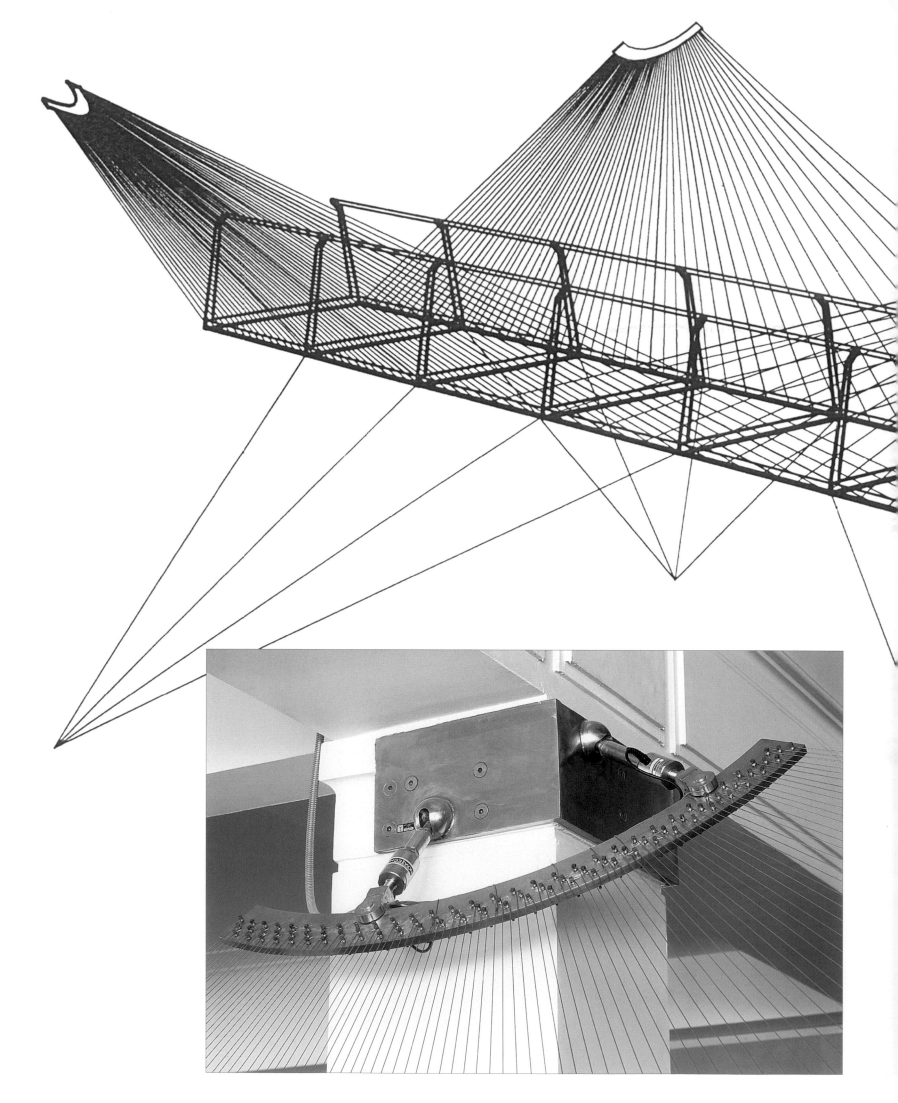

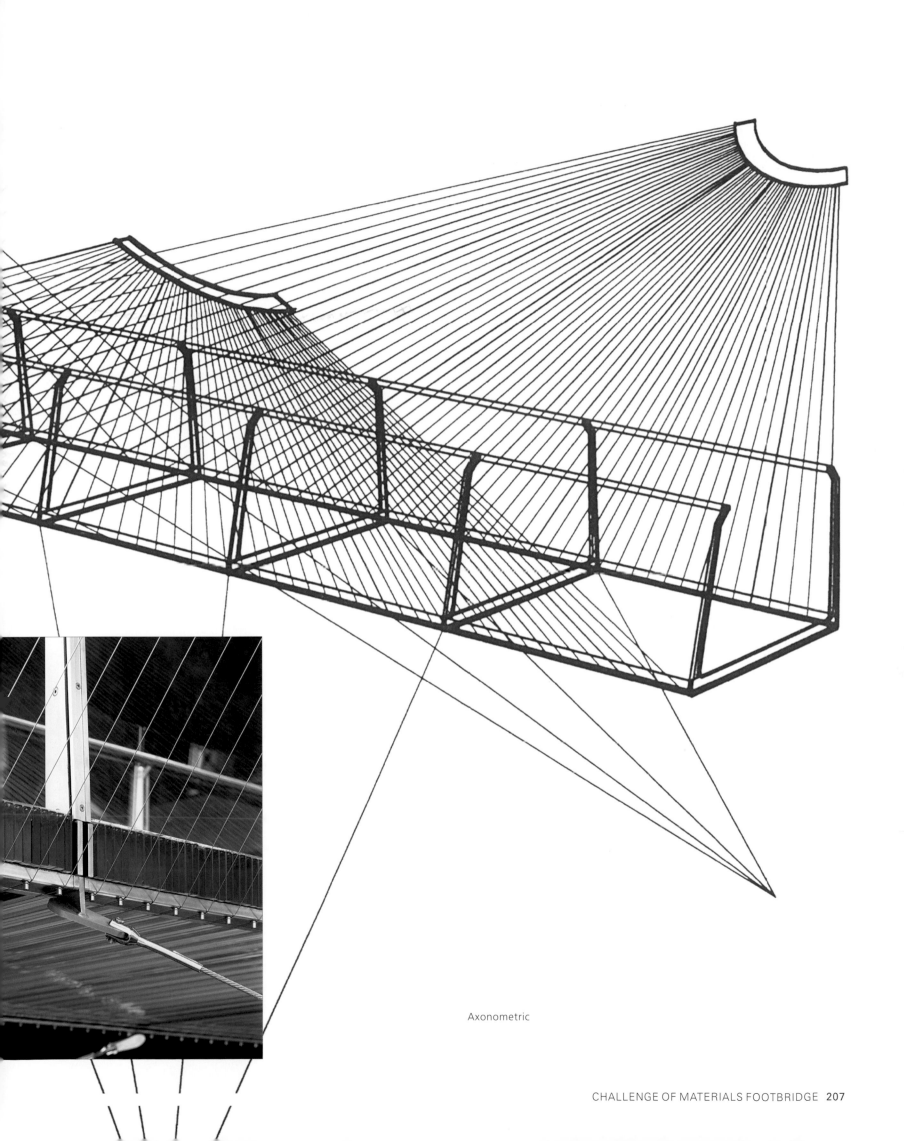

Axonometric

Concept sketches

GATESHEAD MILLENNIUM BRIDGE

Gateshead, England

This project, for a major new crossing over the River Tyne, is part-funded by the National Lottery, and links the newly developed Newcastle Quayside with ambitious plans for the redevelopment of East Gateshead.

The brief required that a clear channel for shipping was retained, whilst maintaining a low-level crossing for pedestrians and cyclists. It was also essential to acknowledge the contextual importance of a place characterised by its bridges.

The opening motion of the design is both its generator and its feature. Bridges that open offer a spectacle, yet are rarely spectacular. This bridge in contrast has visual daring and elegance in its closed position, giving way to theatre and power while in operation.

The idea is one of elegant simplicity. A pair of arches, one forming the deck, the other supporting it, pivots around its mutual springing point to allow shipping business to pass beneath. The motion is efficient and rational, yet dramatic beyond the capabilities of previously explored opening mechanisms. The whole bridge tilts, and in the process the composition undergoes a metamorphosis into a grand arch of great width and grace, in an operation which evokes the action of a closed eye slowly opening.

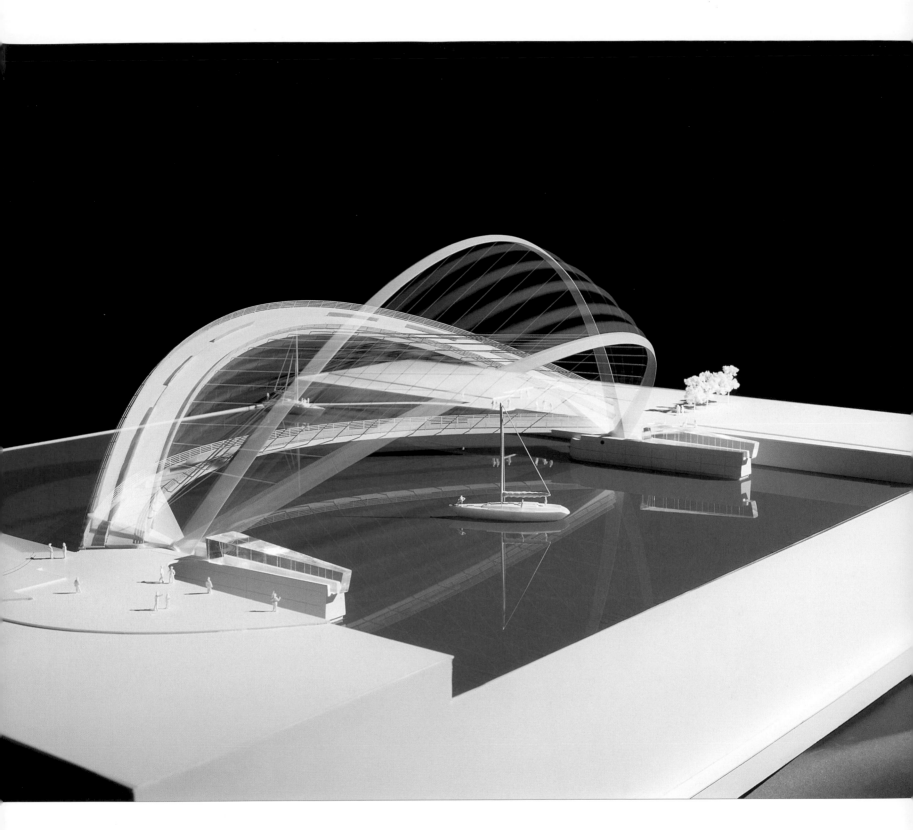

Model showing opening sequence

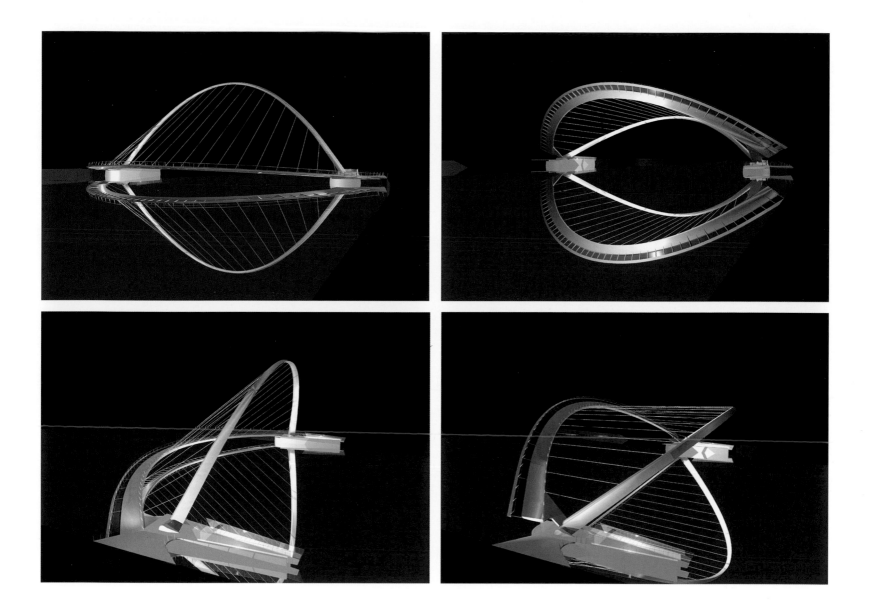

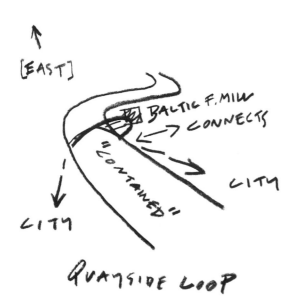

[EAST]

BALTIC F. MILL

CONNECTS

"CONTAINED"

CITY

CITY

QUAYSIDE LOOP

Above Computer model
Left Location sketch

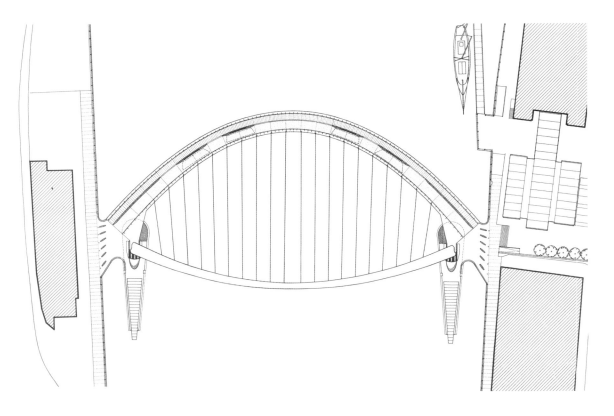

Plan

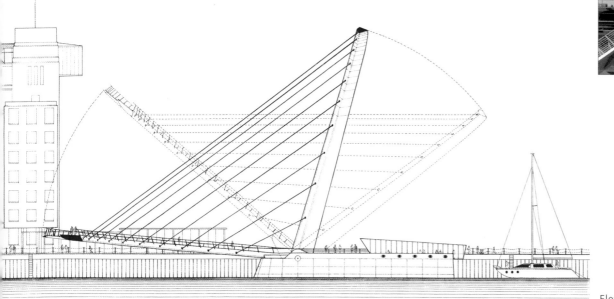

Elevation

LOCKMEADOW FOOTBRIDGE

Maidstone, England

This was the winning entry to a design competition for a new pedestrian bridge across the river Medway, at a sensitive location at the historic core of Maidstone.

The bridge connects the town to new developments on the Lockmeadow floodplain, at a site with specific features fundamental in determining the design of the scheme. The existing backdrop of important medieval buildings and mature trees calls for a minimal and dignified intervention. This is coupled with a critical need to span high over the river and present minimum obstruction to floodwaters. The bridge does not ostentatiously create a landmark, but rather relies on subtler attributes of elegance and efficiency.

The 80m span bridges both the river and the floodplain, bearing to a central pier on the west bank. This pier – the cutwater – constitutes the structural, visual, and functional core of the design. Shaped like the bow of a ship and formed of face finished concrete, the cutwater incorporates steps linking the axial path with the bridge deck. At its head two raking masts flank the steps, providing deck support and lending ceremony to the journey between bridge and river's edge. The deck spans east and west from the cutwater, connecting respectively with the opposing bank and the flood meadow with the lightest of touches.

The steel masts assume a skeletal form, allowing visual penetration and a consequent reduction in profile. Cigar-shaped, and comprising three individual chords separated by triangular spacers, they rake forwards and outwards from the cutwater, their heads central over the deck below. Each span is supported by two pairs of cables, gathering into a single strand at the mast by a series of tributary junctions. The cables anchor to either side of the deck, with pedestrians passing between and under them.

The design employs an extremely shallow deck and a substantially transparent balustrade to provide a minimal profile. The deck, in contrast to conventional techniques, combines both structure and decking as a single element as long aluminium extrusions. A serrated top surface provides a self-finished non-slip traffic surface, while the extrusion below provides a spanning structure with minimal depth.

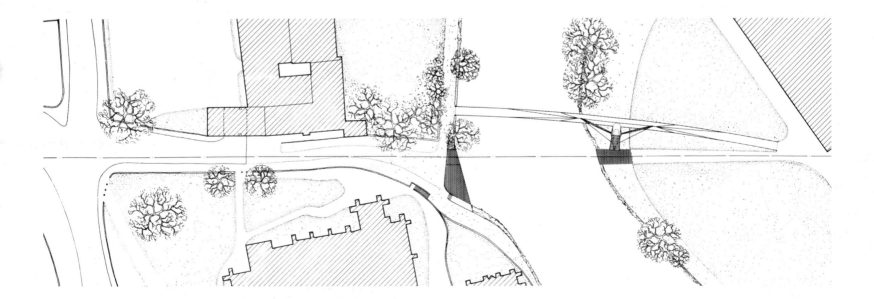

Site plan

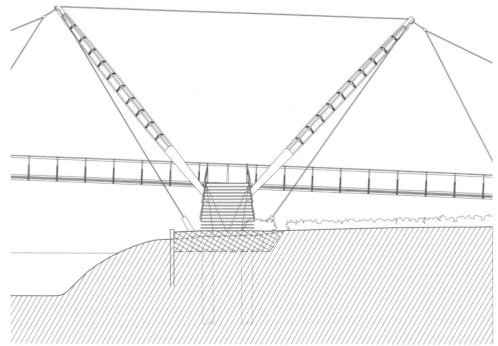

Elevation

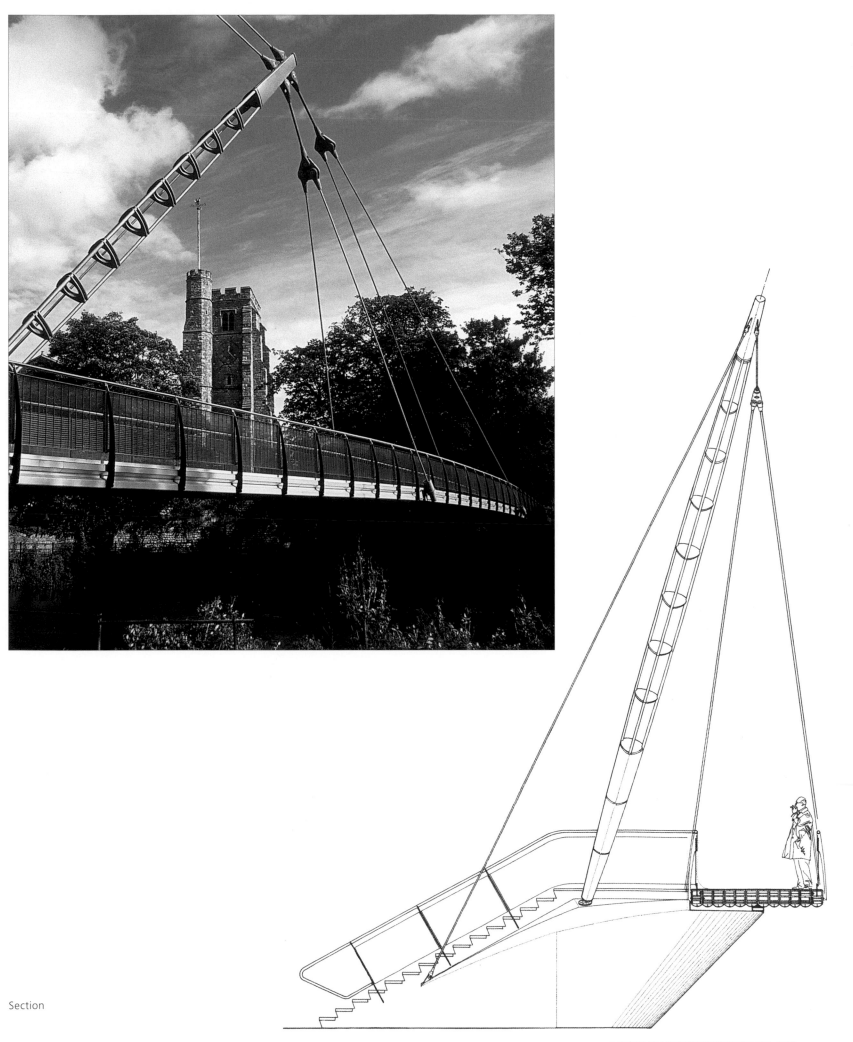

Section

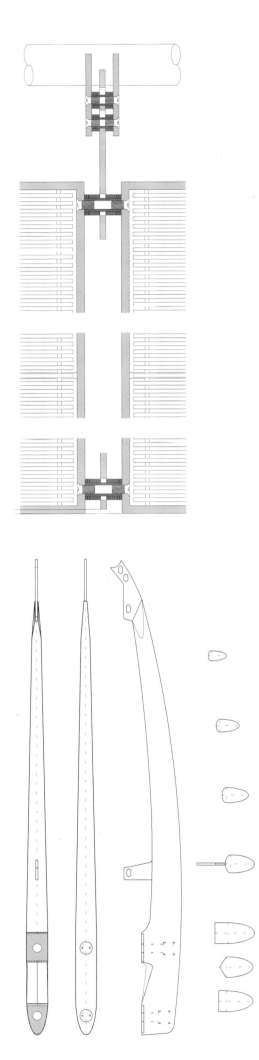

Above right Elevation detail of balustrade
Right Component detail of balustrade

PROJECT INFORMATION

The following information has been collated from material provided by the architects. All costs should be taken as a guide only.

Anthony Hunt Associates with Future Systems Architects
West India Quay Floating Bridge, London, England
Client: London Docklands Development Corporation
Structural engineer: Anthony Hunt Associates (London office)
Main / steelwork contractor: Littlehampton Welding
Quantity surveyor: Bucknall Austin
Project value: £1.7 million
Completed: 1996

Anthony Hunt Associates with John McAslan & Partners
Kelvin Link Bridge, Glasgow, Scotland
Client: University of Glasgow
Structural engineer: Anthony Hunt Associates (Cirencester office)
Quantity surveyors: Davis Langdon & Everest (Glasgow office)
Completion: anticipated 2002

Austin Winkley & Associates
Bridge at the World Association of Christian Communication, London, England
Design team: Tim Gough (architect, cost control & planning supervisor)
Client: World Association for Christian Communication (WACC)
Structural engineer: Atelier One
Main contractor: Hall & Tawse City, Enfield, Middlesex
Main subcontractors/suppliers: John Walker Fabrications, Harlow, Essex (steelwork); GFC Lighting, London (light fittings)
Quantity surveyor: Austin Winkley & Associates
Project value: £45,000
Completed: 1996

Birds Portchmouth Russum Architects
Plashet School Footbridge, Newham, London, England
Client: Education Department, London Borough of Newham
Structural engineer: Techniker
Main contractor: C. Spencers Ltd
Main subcontractors/suppliers: MSI Ltd (steelwork); Architen (fabric)
Quantity surveyor: Gardiner & Theobald
Project value: £530,000
Completed: September 2000

Brookes Stacey Randall
Cardiff Bridges, Cardiff, Wales
Design team: Andrew Fursdon, Matthew Bedward, Nik Bandall, Michael Stacey
Client: Cardiff Bay Development Corporation
Engineer: Atelier One
Main contractor: Gerald Davies
Fabricator: Hayes Engineering
Project value: £125,000
Completed: 1995

Clash Associates
Three Mills Bridge, London, England
Clients: London Industrial; Stratford Development Partnership
Structural engineer: Marks, Healey & Brothwell
Artist: Peter Fink
Quantity surveyors: GHP; Gardiner & Theobald
Completed: 1997

CZWG
Green Bridge, Mile End Park, London, England
Design team: Tibbalds Monroe
Clients: East London Business Alliance; The Environment Trust; London Borough of Tower Hamlets
Structural engineer: Mott MacDonald (ex Bingham Cotterell)
Main contractor: Fitzpatrick Contractors Ltd
Mechanical & electrical engineering: Fulcrum Consulting
Landscape architects: LBTH Landscape
Project manager: Brian Cheetham Partnership
Quantity surveyor: Chandler KBS
Project value: £5.8 million
Completed: June 2000

Donaldson + Warn
Tree Top Walk, Valley of the Giants, Walpole, Nornalup National Park, Western Australia
Design team: Geoff Warn (Design Director), Jane Bennetts (Project Architect), Matthew Crawford, Graham Christ, Martyn Hook, Andrew Scafe, John Sunderland, David Jones (Environmental artist), Tracy Churchill (CALM)
Client: Western Australian Department of Conservation and Land Management (CALM)
Structural engineer: Adrian Roberts, Ove Arup & Partners
Quantity surveyor: Ralph and Beattie Bosworth
Land surveyor: Fowler Surveys
Hydraulic consultant: Harding & MacDonald
Main contractor: Future Engineering & Communication
Main subcontractors/suppliers: Folkes Smith & Associates (pylon supply)
Project value: $ (Australian) 2 million
Completed: August 1996
Awards: Western Australian Civic Design Award: Premiers Awards, 1996; Western Australian Civic
Design Award: Specific Feature Award, 1996; Australian Institute of Landscape Architecture National
Project Awards: Design Category, 1996; Royal Australian Institute of Architect (WA Chapter) Award of Merit for Civic Design, 1997; Royal Australian Institute of Architects National Architecture Awards: Citation for Access, 1997; BHP Australian Steel Award for Architecture, 1997; Western Australian Tourism Award, 1998

Eva Jiricna Architects
Koliste Footbridge, Brno, Czech Republic
Design team: Eva Jiricna, Duncan Webster, Geoff Whittaker, Adrian Welch
Client: Unistav A.S.
Structural engineer: Techniker
Main contractor: Olomouc Metalworkers
Project value: £600,000
Completed: 1998

Foster & Partners
Millennium Bridge, London, England
Design team: Norman Foster (Principle), Malcolm Reading (Project Director), Ken Shuttleworth, Andy Bow, Catherine Ramsden, Jason Salero (all of Foster and Partners) with Sir Anthony Caro and Ove Arup & Partners
Client: Millennium Bridge Trust, London Borough of Southwark
Structural & mechanical engineer: Ove Arup & Partners
Main contractor: McAlpine & Monberg Thorsen Joint Venture
Lighting design: Claude Engle; Ove Arup & Partners
Consultants: Sir Anthony Caro (sculptor); Space Syntax
Quantity surveyor: Davis Langdon & Everest
Project value: £14 million
Completed: 2000

Giancarlo De Carlo
Dogana Gate, Republic of San Marino
Design team: Paolo Castiglioni, Antonio Troisi, Giuseppe Carniello, Takaoki Kanehara
Client: Republic of San Marino
Structural engineer: Giuseppe Carniello
Lighting design: Piero Castiglioni
Glass design: Anna De Carlo, Giancarlo De Carlo, Daniela Poletti
Models: Ornella Calatroni, Takaoki Kanehara
Completed: 1996

Günther Domenig
Footbridge over the River Mur, Graz, Austria
Design team: Gunther Domenig and Hermann Eisenköck with Gerhard Wallner and Johannes Dullnigg
Client: City of Graz
Structural engineer: Professor Harald Egger
Main contractor: VOEST Alpine MCE, Linz (steel construction)
Main subcontractors/suppliers: ALPINE Bau, Graz (foundation); AEG (lighting)
Project value: ATS 11 million
Completed: November 1992

Hodder Associates
Corporation Street Footbridge, Manchester, England
Design team: Stephen Hodder, Peter Williams, Helen Roberts, Stewart Jones
Client: Manchester Millennium-
Structural engineer: Ove Arup & Partners
Main contractor: Bovis
Main subcontractors/suppliers: Watson Steel (structural steelwork)
Quantity surveyor: Poole Stokes Wood
Project value: £500,000
Completed: November 1999

Howley Harrington Architects
Liffey Pedestrian Bridge, Dublin, Republic of Ireland
Project Architect: Seán Harrington
Client: Dublin Corporation
Structural engineer: Price & Myers Consulting Engineers, London
Main contractor: Ascon Ltd
Main subcontractors/suppliers: Thompson Engineering Ltd, Carlow (steel truss fabrication);

Banagher Concrete Ltd (concrete shell abutments)
Lighting design: Howley Harrington Architects in conjunction with Lighting Design Partnership
Quantity surveyor: Austin Reddy & Co. Ltd
Project value: IR£1.6 million
Completed: December 1999

Ian Ritchie Architects
Natural History Museum Ecology Galleries, London, England
Design team: Ian Ritchie, Henning Rambow, Klaus Schnedkamp, Edmund Wan
Client: Natural History Museum
Structural engineer: Arup (Peter Rice and Sarah Meldrum)
Main contractor: Walter Lawrence (City and Southern) Ltd
Main subcontractors/suppliers: The Thanet Foundry and Engineering Co. Ltd (steelwork); Pegasus (woodwork); Twide (glazing)
Quantity surveyor: Perspective Project Management Ltd
Project value: £1.5 million (including exhibition fit-out)
Completed: March 1991

Jiri Strasky
Pedestrian Bridge across the Rogue River, Grants Pass, Oregon, USA
Design team: Jiri Strasky, Consulting Engineer; Gary Rayor, OBEC, Consulting Engineer
Client: City of Grants Pass, Oregon
Structural engineer: OBEC, Consulting Engineer
Main contractor: Holm II Construction, Stayton, Oregon
Main subcontractors/suppliers: Avar, Campbell, California (post tensioning); Morse Brothers Prestress, Harrisburg, Oregon (precastor)
Quantity surveyor: OBEC, Consulting Engineer
Project value: $1.32 million
Completed: summer 2000

Jiri Strasky
Sacramento River Trail Pedestrian Bridge, Redding, California, USA
Design team: Jiri Strasky, Consulting Engineer; Charles Redfield, Consulting Engineer
Client: City of Redding
Structural engineer: Charles Redfield, Consulting Engineer
Main contractor: Shasta Constructors, Redding
Main subcontractors/suppliers: Avar, Campbell, California (post tensioning)
Quantity surveyor: Charles Redfield, Consulting Engineer
Project value: $0.60 million
Completed: spring 1990

Jiri Strasky
Vranov Lake Pedestrian Bridge, Czech Republic
Design team: Jiri Strasky, Ilja Husty, SHP Consulting Engineers
Client: City of Vranov
Structural engineer: Dopravni stavby, Projekce
Main contractor: Dopravni stavby & Mosty
Main subcontractors/suppliers: Geotest Brno (rock anchors)
Quantity surveyor: Dopravni stavby, Projekce

Project value: $1 million
Completed: spring 1993

Jiri Strasky
Willamette River Pedestrian Bridge, Eugene, Oregon, USA
Design team: Jiri Strasky, Consulting Engineer; Gary Rayor, OBEC, Consulting Engineer
Client: City of Eugene, Oregon
Structural engineer: OBEC, Consulting Engineer
Main contractor: Mowat Construction, Vancouver, WA
Main subcontractors/suppliers: Dywidag-Systems International, Long Beach, California (post tensioning subcontractor); Eugene Sand and Gravel, Eugene, Oregon (precastor)
Quantity surveyor: OBEC, Consulting Engineer
Project value: $2.40 million
Completed: spring 1999

Judah
Linkbridge 2000, London, England
Design team: Gerry Judah (lead designer), Marcel Ridyard (detailed design)
Clients: Sustrans; Woolwich Development Agency
Structural engineer: Ingealtral Main contractor: Greenwich Council
Main subcontractors/suppliers: Code-Arc, Ipswich (metal); Broadmead Cast Stone, Maidstone (stonework)
Project value: £100,000
Completed: April 2000

Lifschutz Davidson Ltd
Hungerford Bridge Millennium Project, London, England
Design team: WAP Group; Speirs & Major (lighting design)
Client: Westminster City Council
Structural engineer: WSP Group
Quantity surveyor: Davis Langdon & Everest
Project value: £40 million
Completed: 2002

Lifschutz Davidson
Royal Victoria Dock Bridge Project, London, England
Design team: Allott & Lomax; Equation Lighting Design
Client: London Docklands Development Corporation
Structural engineer: Techniker
Main contractor: Kier London
Quantity surveyor: Davis Langdon & Everest
Project value: £5 million
Completed: 1999
Awards: AJ/Bovis Awards – Royal Academy, 1996

Marc Mimram
Solferino Footbridge, Paris, France
Design team: Marc Mimram (architect) with Marc Mimram Ingenierie (engineers)
Client: EPMOTC (Etablissement Public de Maîtrise d'Ouvrage des Travaux Culturels)
Structural engineer: Marc Mimram
Completed: 2000

Mark Lovell Design Engineers
Black Dog Hill Bridge, near Calne, Wiltshire, England
Design team: Mark Lovell Design Engineers
Client: North Wiltshire District Council
Structural engineer: Mark Lovell Design Engineers

Main subcontractors/suppliers: Conley Structural Timbers; Denning Rudman & Bent; ECC Timber
Quantity surveyor: John Cox
Project value: £225,000
Completed: December 1999

Napper Architects
River Irthing Footbridge, Hadrian's Wall, Cumbria, England
Designer: Christopher Rainford
Client: The Countryside Agency with Cumbria County Council
Structural engineer: Ove Arup & Partners, Newcastle
Main contractor: John Laing Construction Ltd, Newcastle
Main subcontractors/suppliers: ASD Glen Metals (steel supply); Hartlepool Erection Co. Ltd (steel manufacturing/erection); Sarum Hardwood Structures Ltd (timber decking); Expanded Metal Company Ltd (tread finish)
Quantity surveyor: n/a
Project value: £160,000
Completed: January 1999
Awards: RIBA Architectural Award, 1999; Robert Stephenson Engineering Award, 1999; Royal Fine Art Commission / BSkyB Building of the Year Award (Bridges), 1999; Civic Trust Award, 2001

Nicholas Lacey & Partners
Rotherhithe Tunnel Bridge, London, England
Client: London Docklands Development Corporation
Structural engineer: Bryn Bird of Whitby & Bird Engineers (concept); W.S. Atkins Engineers (executive)
Project value: £550,000
Completed: 1998

Nicholas Lacey & Partners
St Saviour's Dock Footbridge, London, England
Client: London Docklands Development Corporation
Structural engineer: Bryn Bird of Whitby & Bird Engineers
Main contractor: Christian & Nielson
Main subcontractors/suppliers: Littlehampton Welding
Project value: £750,000
Completed: 1996
Awards: Civic Trust Award, 1998; RIBA London Region Award for Architecture, 1997

Office for Metropolitan Architecture
Museum Park, Rotterdam, The Netherlands
Design team: Yves Brunier, Petra Blaisse, Rem Koolhaas with Tony Adam, Maartje Lammers, Gregor Mescherowsky
Client: City of Rotterdam
Project value: Dfl 4 million
Completed: 1994

Pei Cobb Freed & Partners
Miho Museum, Shigaraki, Shiga Prefecture, Japan
Design team: I.M. Pei, Perry Chin, Tim Culbert, Chris Rand, Carol Averill, Price Harrison, Celia Imrey, Hubert Poole
Client: Shumei Culture Foundation
Structural consultants: Leslie E. Robertson Associates; Aoki Structural Engineers; Nakata & Associates; Whole Force Studio

Mechanical consultants: P.T. Morimura, Makoto Kanai, Takao Kawauchi, Shizu Takazawa
Main contractor: Shimuzu Corporation, Osaka (general); Takasago Netsugaku Co. & Sugakogyo Co. (mechanical); Kandenko (electrical
Main subcontractors/suppliers: Y.K. Kap (skylight curtain wall and sunshading); Tajima (vertical glass wall sash); Sumitomo: Heavy Industry (space frame); Toyo
Precast: Concrete Corporation (concrete); Sekigahara Stone Corporation (stone); Okuju (metals); Nittobo (interior walls); Miyazaki (wood); Glasbau Hahn & Kokuo (exhibit showcases); Kohseki (garden works)
Lighting: Fisher Marantz Renfroe Stone
Completed: 1997

RFR and Feichtinger Architectes
Bercy-Tolbiac Footbridge, Paris, France
Design team: Feichtinger Architectes (Dietmar Feichtinger, architect; Barbara Feichtinger-Felber; Bernardo Bader; Christian Pilcher; Marta Mendonça; Guy Deshayes; José-Luis Fuentes; Montse Ferres; Mario Lins; Jo Pesendorfer (site photographs), Woytek Sepiol (model photographs); Eddie Young (renderings); RFR (engineers)
Client: Ville de Paris
Structural engineer: RFR
Completion: due 2002

RFR and Schulitz & Partners
Skywalk, Hanover, Germany
Design team: RFR (engineers), Schulitz & Partners (architects)
Client: Deutsche Messe AG
Structural engineer: RFR
Completed: 1998

RFR and Kisho Kurokawa
Japan Bridge, Paris, France
Design team: RFR (engineers), Kisho Kurokawa (architect)
Client: SARI Construction for SCI Puteaux Kupka
Structural engineer: RFR
Construction manager: SARI Ingenierie
Project value: FF 28 million
Completed: 1993
Awards: Concours des Plus Beaux Ouvrages de la Construction Métallique, 1994

Schlaich Bergermann & Partner
Convertible Bridge over the Inner Harbour, Duisburg, Germany
Design team: Jörg Schlaich, Michael Stein, Peter Schulze
Client: Innenhafen Duisburg Entwicklungsgesellschaft
Structural engineer: Schlaich Bergermann & Partner
Main contractors: Stahlbau Raulf, Duisburg; Jaeschke und Preuss, Duisburg
Project value: DM 5.5 million
Completed: 1999

Schlaich Bergermann & Partner
Folding Bridge over the Förde, Kiel, Germany
Design team: Jörg Schlaich, Jan Knippers, Volkwin Marg, Reiner Schröder of Schlaich Bergermann with von Gerkan, Marg & Partner, Hamburg
Client: Landeshauptstadt Kiel

Structural engineer: Schlaich Bergermann & Partner
Main contractors: Neptun Stahl Objektbau, Rostock
Project value: DM 10 million
Completed: 1997

Schlaich Bergermann & Partner
Pedestrian Bridge over the Neckar River, near Max-Eyth-See, Stuttgart, Germany
Design team: Jörg Schlaich, Hans Schober, Thomas Fackler, Johen Bettermann of Schlaich Bergermann with Brigitte Schlaich-Peterhans, architect
Client: Landeshauptstadt Stuttgart
Structural engineer: Schlaich Bergermann & Partner
Main contractor: Wayss & Freitag, Stuttgart; Pfeifer, Memmingen
Project value: DM 4.1 million
Completed: 1988

Schlaich Bergermann & Partner
Pedestrian Bridge over the Weser River, Minden, Germany
Design team: Jörg Schlaich, Andreas Keil
Client: Stadt Minden
Structural engineer: Schlaich Bergermann & Partner
Project value: DM 4.5 million
Completed: 1996

Whitby Bird & Partners
Oracle Bridges, Reading, England
Design team: Richard Jobson, Des Mairs, Simon Skeffington, Jane Sheridan
Client: Hammerson plc
Structural engineer: Whitby Bird & Partners
Main contractor: Norwest Holst
Main subcontractors/suppliers: Littlehampton Welding
Quantity surveyor: C. Sweet
Project value: £700,000
Completed: November 1999
Awards: British Construction Industry Awards, 2000

Whitby Bird & Partners
River Lune Millennium Bridge, Lancaster, England
Design team: Mark Whitby, Richard Jobson, Des Mairs, Tamsin Ford, Stephen Boyd
Client: Lancaster City Council
Structural engineer: Whitby Bird & Partners
Main contractor: Henry Boot
Main subcontractors/suppliers: Kent Structural Marine
Quantity surveyor: Davis Langdon & Everest
Project value: £1.8 million
Completed: February 2001

Whitby Bird & Partners
Shanks Millennium Bridge, Peterborough, England
Design team: Richard Jobson, Des Mairs, Tamsin Ford, Stephen Foster
Client: Peterborough Environment City Trust
Structural engineer: Whitby Bird & Partners
Main contractor: May Guerney
Main subcontractors/suppliers: Fairfield-Mabey
Quantity surveyor: Posford Haskoning
Project value: £750,000
Completed: December 2000
Awards: Royal Fine Art Commission/BSkyB Building of

the Year (bridge category), 2001; Structural Steel Design Award commendation, 2001

Whitby Bird & Partners
York Millennium Bridge, York, England
Design team: Mark Whitby, Richard Jobson, Des Mairs, Scott Lomax
Client: York Millennium Bridge Trust
Structural engineer: Whitby Bird & Partners
Main contractor: C. Spencer Ltd, Barrow on Humber
Contractor's designer: Bennett Associates, Rotherham
Fabricator: Meldan Fabrications Ltd, Barrow on Humber
Quantity surveyor: Davis Langdon & Everest
Landscape architect: Robert Rumney Associates, Sevenoaks, Kent
Project value: £2.2 million
Completed: April 2001

Wilkinson Eyre Architects
Butterfly Bridge, Bedford, Bedfordshire, England
Design team: Keith Brownie, Jim Eyre, James Parkin, Geoff Turner
Client: Bedford Borough Council
Structural engineer: Jan Bobrowski & Partners, Twickenham
Main contractor: Littlehampton Welding Limited
Quantity surveyor: Davis Langdon & Everest
Project value: £375,000
Completed: May 1997

Wilkinson Eyre Architects
Challenge of Materials Footbridge, London, England
Client: Science Museum London
Structural engineer: Whitby Bird & Partners
Main contractor: Hubbard Architectural Metalwork Limited
Sound and light artist: Ron Geesin
Project value: £200,000
Completed: May 1997
Awards: Glassex Industry Awards Finalist, 1998; awarded 'Millennium Product' status by the Design Council, 1998

Wilkinson Eyre Architects
Gateshead Millennium Bridge, Gateshead, Tyne and Wear, England
Client: Gateshead Metropolitan Council
Structural engineer: Gifford & Partners
Main contractors: Harbour & General; Volker Stevin
Project value: £22m
Completed: July 2001
Awards: Royal Academy AJ/Bovis Grand Award, 1997

Wilkinson Eyre Architects
Lockmeadow Footbridge, Maidstone, Kent, England
Client: Maidstone Borough Council (supported by the National Lottery as a Millennium Project)
Structural engineer: Flint & Neill Partnership
Main contractor: Harbour & General/Volker Stevin
Project value: £650,000
Completed: October 1999
Awards: Structural Achievement Award, 2000; Institute of Civil Engineers Merit Award, 2000